Residential, Industrial, and Institutional Pest Control

Patrick J. Marer
Pesticide Training Coordinator
IPM Education and Publications
University of California, Davis

TECHNICAL EDITORS:
Mary Louise Flint
Director, IPM Education and Publications
University of California, Davis

Michael K. Rust
Professor of Entomology
University of California, Riverside

University of California
Statewide Integrated Pest Management Project
Division of Agriculture and Natural Resources
Publication 3334
Version 3334

1991

ORDERING

For information about ordering this publication, write to:

> Publications
> Division of Agriculture and Natural Resources
> University of California
> 6701 San Pablo Avenue
> Oakland, California 94608-1239

or telephone (510) 642-2431

> Publication #3334

ISBN 0-931876-93-1

Acknowledgments

This manual was produced under the auspices of the University of California Statewide Integrated Pest Management Project through a Memorandum of Understanding between the University of California and the California Department of Food and Agriculture. It was prepared under the direction of Mary Louise Flint, Director, IPM Education and Publications, University of California, Davis, and Michael W. Stimmann, Coordinator, Office of Pesticide Information and Coordination, University of California.

Production

Design and Production Coordination: Seventeenth Street Studios
Photographs: Jack Kelly Clark
Drawings: David Kidd
Editing: Louise Eubanks/Andrew Alden

Principal Contributors and Reviewers:

The following people were key resources for specific subject matter areas. They provided ideas, information, and suggestions and reviewed the many manuscript drafts.

M. Flint, University of California, Davis
J. Glenn, University of California, Davis
M. Hurlbert, University of California, Berkeley
C. Koehler, University of California, Berkeley
J. Meyers, University of California, Riverside
J. Munro, Pest Control Operators of California, West Sacramento
G. Okumura, Okumura Biological Institute, Sacramento
W. Olkowski, BIRC, Berkeley
M. Rust, University of California, Riverside
T. Salmon, University of California, DANR-North Region, Davis
A. Slater, University of California, Berkeley
E. Soderstrom, USDA-ARS, Fresno
M. Stimmann, University of California, Davis

Special Thanks

The following persons have generously provided information, offered suggestions, reviewed draft manuscripts, assisted in setting up photographs, or helped in other ways.

R. Baker, California State Polytechnic University, Pomona
R. Barrett, Bugman Pest Control, Inc.
J. Bean, Dow Chemical Company
P. Bone, University of California, Cooperative Extension
M. Brush, University of California, Davis
W. Chaney, University of California, Cooperative Extension
W. Dost, University of California, Berkeley
W. Ebeling, University of California, Los Angeles
H. Elting-Ballard
M. Feletto, Cal-OSHA Consulting Service
M. Ferreira, California Structural Pest Control Board
D. Gibbons, California Department of Food and Agriculture
C. Joshel, University of California, Davis
R. Krieger, California Department of Food and Agriculture
G. Kund, University of California, Riverside
M. Lawton, Western Exterminator Company
C. Levine, University of California, Riverside
R. Molinar, University of California, Cooperative Extension
D. Mueller, Fumigation Service and Supply, Inc.
E. Mussen, University of California, Davis
W. Newell, Dixon, California
L. Perot, California Department of Food and Agriculture
E. Perry, University of California, Cooperative Extension
D. Rierson, University of California, Riverside
S. Segal, U.S. Environmental Protection Agency
J. Smith, University of California, Riverside
R. Smith, University of Arizona
F. Stegmiller, Harold, California
W. Steinke, University of California, Davis
L. Strand, University of California, Davis
M. Takeda, California Department of Food and Agriculture
R. Yamaichi, University of California, Davis
M. Zavala, University of California, Davis

Contents

1 Introduction

Many living organisms can be pests in or around structures. These may be insects and related arthropods; others include fungi, weeds, rats, mice, bats, and certain birds. Some pests, such as those that damage structural wood, furnishings, or fabrics or pests that infest or contaminate stored food and other items, cause huge economic losses. A few pests spread disease organisms that can cause serious human illness. Certain types of pests inflict annoying or painful stings or bites. In addition, nuisance pests such as earwigs or sowbugs are unwelcome invaders in homes and can also contaminate products or cause legal concerns in commercial areas.

Pests that damage property, injure people, or affect people's quality of life need to be managed. However, management methods should be safe, effective, and economical. Sometimes there are several ways available to control pests, such as sanitation and habitat modification, trapping, and the use of pesticides.

Pesticides are usually very effective tools for controlling pests. But if you use pesticides improperly, you may injure yourself and other people and may create environmental problems. In addition, some improperly used pesticides may damage treated surfaces. To prevent problems or accidents, you must always follow pesticide label instructions and use basic common sense. This book contains important information to help you use pesticides properly in and around structures.

How to Use this Book

Use the information in this book to make effective pest management decisions that will reduce hazards to yourself, other people, and the environment. If you are preparing for the *Residential, Industrial, and Institutional* pesticide applicator examinations, use this book as your primary study guide.

When you apply pesticides in structures, use this book as a reference for information on pests, pest management, and pesticide use. You should find the book helpful if you supervise or train persons who handle or apply pesticides in any type of residential, industrial, or institutional situation.

Helpful information about pesticides and their alternatives is included in shaded boxes such as the one on page 3. Many tables are also included to summarize important points made in the text and to provide additional information.

This is the second volume of a series of manuals developed expressly to teach safe and correct ways to use pesticides. Volume 1 of the Pesticide Application Compendium series, *The Safe and Effective Use of Pesticides*, gives general information on pest identification and management, pesticides, pesticide safety, and pesticide application equipment.

The references at the end of this book include many well-illustrated books and pamphlets that provide additional information on identification, biology, and management of pests. Furthermore, you can obtain pest management information and control recommendations for specific pests from advisors in University of California Cooperative Extension offices located in most California counties.

Pesticide Concerns

Pesticides have been and will continue to be the focus of a great deal of public attention. Many people have genuine concerns over the use of pesticides, although many others use them regularly to manage pest problems and find the risks acceptable.

Pesticide concerns arise from reported incidents where exposure has produced mild to severe illness (or death) in farmworkers, pesticide applicators, manufacturing plant workers, and even consumers of improperly treated produce.

Injuries caused by some pesticides have included skin rashes, headache, nausea, and nervous system disorders. Long-term or chronic illnesses associated with or suspected of being caused by certain pesticides include cancer, birth defects, and reproductive disorders.

Pesticides have also been implicated in environmental problems such as groundwater contamination and wildlife injury.

The debate over the advisability of using pesticides will continue for many years. In the meantime, pesticide hazards must be reduced by proper handling and application techniques, accurate timing of applications, and by seeking and using alternate control methods whenever possible.

FIGURE 1-1

Pest management in structures is regulated by the California Department of Food and Agriculture and the California Department of Consumer Affairs, Structural Pest Control Board.

Regulatory Agencies

Two California state agencies establish qualifications, administer examinations, and license or certify persons engaged in residential, industrial, and institutional pest control (Figure 1-1). These are the Department of Food and Agriculture and the Structural Pest Control Board. Persons using pesticides, other than fumigants, in their own homes are not required to be licensed. Legally, however, they must use all pesticides according to label instructions; they cannot apply certain *restricted-use* products without becoming certified by the California Department of Food and Agriculture (CDFA).

California Department of Food and Agriculture

The CDFA certifies people who apply or supervise the application of restricted-use pesticides in residential, industrial, or institutional buildings (and other areas) as part of their regular employment. Such people include building superintendents, caretakers, and maintenance workers who are employed by apartment owners, schools, government agencies, manufacturing plants, private businesses, hospitals, or similar facilities. (In California, structural pest control work for hire can only be performed by a person holding a valid structural pest control license issued by the Structural Pest Control Board—see below.)

Pest control businesses that fumigate stored agricultural products, perform other types of post-harvest pest control on agricultural products, conduct weed control in or around buildings (separate from landscape maintenance), or are involved in the preservation or preventive treatment of wood products for hire must have an agricultural pest control business license. They must also employ a CDFA-qualified applicator in the Residential, Industrial, and Institutional category to perform or supervise pesticide applications and supervise the pest control operations of the business.

The CDFA regulates all pesticide use within California, issues permits for restricted-use pesticides, and enforces regulations dealing with pesticide safety, handling, application, and disposal. Permits and enforcement functions of the CDFA are coordinated primarily through county agricultural commissioners' offices.

Structural Pest Control Board

The Structural Pest Control Board of the California Department of Consumer Affairs administers examinations and issues licenses to individuals performing any type of structural pest control for hire, including pesticide application. The Board establishes the state's structural pest control standards and training requirements. It also regulates and monitors the pest control activities of businesses and individuals. Enforcement functions of the Structural Pest Control Board are administered through county agricultural commissioners' offices.

Structural Pest Control Board licenses are issued in three different categories: *Branch 1*—fumigation; *Branch 2*—general pests; and *Branch 3*—wood-destroying pests and organisms. Two levels of competence—*Operator* and *Field Representative*—are recognized in each branch. An approved program of training and documented work experience must be followed by passing written examinations to qualify for structural pest control licenses. Special training courses are offered through public and private organizations. Some pest control businesses in California are approved to train their employees to qualify them for licensing.

License and Certificate Renewal

Licenses and certificates must be periodically renewed by the issuing agency. Renewal requires payment of fees and completion of a specified number of hours or points of approved continuing education. People may also retake the appropriate examinations to renew their license or certificate.

Contact the agencies listed in Table 1-1 for information on training requirements, examinations, licensing, and continuing education requirements.

TABLE 1-1

Agencies Regulating Residential, Industrial, and Institutional Pesticide Applications.

STRUCTURAL PEST CONTROL BUSINESSES AND THEIR EMPLOYEES

For educational requirements, a listing of approved training institutions, and licensing and examination information, contact:

Department of Consumer Affairs
Structural Pest Control Board
1430 Howe Avenue
Sacramento, California 95825
(916) 924-2291

Three categories of licensing are provided:

Branch 1: Fumigation

Branch 2: General Pests

Branch 3: Wood-Destroying Pests and Organisms

AGRICULTURAL PEST CONTROL BUSINESSES AND THEIR EMPLOYEES

For licensing and examination information, contact:

Department of Food and Agriculture
Pesticide Enforcement
1220 N Street, Room A170
Sacramento, California 95814
(916) 322-5032

OTHER APPLICATORS

Employees of apartments, hospitals, schools, and so on who apply pesticides or supervise pesticide application as part of their employment need a CDFA Qualified Pesticide Applicator License or Certificate in the residential, industrial, and institutional category. For information on training requirements and to apply for examination, contact:

Department of Food and Agriculture
Pesticide Enforcement
1220 N Street, Room A170
Sacramento, California 95814
(916) 322-5032

Two examinations are required, one covering pesticide laws and general aspects of pesticide use and application, and the other specifically for residential, industrial, and institutional pesticide use. Volume 1 of the Pesticide Application Compendium is a study guide for general aspects of pesticide use, and this volume is a study guide for residential, industrial, and institutional pesticide operators.

2 Pest Management

Pest management involves safely preventing, reducing, or eliminating unwanted organisms. To do this, you must learn about the habits and life cycles of many pests and understand the conditions that affect pest populations. A good pest management program follows the principles of integrated pest management (IPM).

One important pest management practice commonly used around structures is prevention of pest problems. Managing pests through *prevention* is usually less expensive than trying to control a pest population that has already become established. Furthermore, pest prevention reduces the chance for substantial economic loss or irreversible damage. Prevention avoids some of

Integrated Pest Management

Integrated pest management is an ecological approach to managing weeds, insects, vertebrates, and other pest organisms that often provides economical, long-term protection from pest damage. IPM has been shown to be very successful in commercial agricultural situations. The benefits of IPM are becoming more widely recognized in the management of urban pest problems.

Pests must be properly identified so aspects of their life cycle and developmental stages can be understood and so their activity can be monitored. Conditions that promote or support the pest are identified so they can be either eliminated or suppressed.

Management methods are appropriate to the life cycle and development stages of the pest. Usually, two or more management methods are used, and commonly different methods are used at other times or in different locations, rather than using the same method for the same pest at all times.

Control methods that might be used in an IPM program include exclusion, sanitation, modifying or eliminating habitats, biological control, and the selective use of pesticides.

Pesticides usually play a important role in an IPM program. However, pesticides are selected carefully and are nearly always combined with other control methods. The timing of the pesticide application is especially important. Pesticides are selected to be least disruptive to natural controls that may be present. Environmental concerns and human and animal safety are an utmost priority. Sometimes an emphasis is placed on spot treatment or reduced rates of the pesticide so that smaller quantities of pesticide need be applied.

An important component of an IPM program involves frequent evaluation of the control strategies and modification of the approach to keep pace with changes or anticipated changes in the pest's activities.

the disruption associated with control efforts that may be needed after pests become established.

Once a pest becomes established, the most common pest management goal is to *eliminate* it. Elimination can only be successful if the conditions that originally favored the pest can be modified or the pest's entry into the area can be completely blocked.

LOCATING AND MONITORING PESTS

Decisions to use pesticides and other control methods should be based in part on pest detection and monitoring results. Visually inspecting an area where pests or their damage is observed is the most common method of detection. Inspection involves careful and thorough searching in and around a structure for signs of the pest and conditions that favor pest buildup. Monitoring is a systematic method of observing pests or pest signs over a period of time. Monitoring may help you detect unwanted pests and determine where pests are coming from and where they are living. Monitoring is also helpful in evaluating control programs. Special devices and tools are available to detect and monitor certain types of pests.

Visual Inspection

The purpose of a visual inspection is to search for evidence of pests. During an inspection, look for: (1) conditions such as food, shelter, access, and suitable environments that favor pests; (2) signs of pest damage, entry, or presence (such as tracks, trails, droppings, nests, and cast skins); and (3) the pest itself.

When you make an inspection, you may find it helpful to prepare sketches of the structure and surrounding areas (Figure 2-1). Include locations

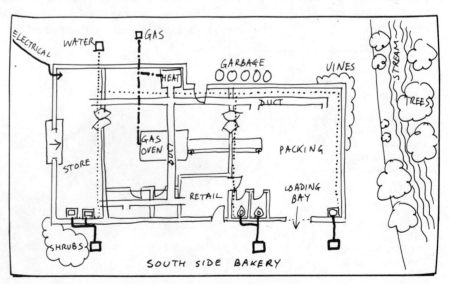

FIGURE 2-1.

Prepare a diagram of the structure and surrounding areas, as shown here, while inspecting for pests and pest damage. Indicate the locations of doors, windows, utilities, ducting, and any other areas that will require special care or attention.

of heating or air conditioning ducts and vents, plumbing inlets, attic, basement, and crawlway vents, wall voids, sub-cabinet voids, and other features of the building which allow pests to get in or which provide shelter for them. Also, observe conditions that may cause problems during pest control operations. Be sure to note areas of poor or faulty construction or places where the building has been damaged by the careless operation of equipment, leaking plumbing, or other reasons. Also, note areas that you were unable to inspect because they were inaccessible. Show the locations of trees, shrubs, trash and garbage storage, water sources, and other features of the surrounding area that may attract or harbor pests or promote pest buildup.

Detection and Monitoring Devices

Different types of simple devices can assist you in detecting and monitoring many of the pests found in structures. These include pheromones and other attractants, light traps, flypaper, spring traps, glue boards, and nontoxic tracking powder.

Pheromones and Other Attractants. Pheromones are chemicals normally produced by certain insects (and other animals) to affect the behavior of individuals of the same species. Pheromones are used by insects for mating, aggregation, feeding, trail following, and recruitment. Synthetically made pheromones mimic the action of pheromones produced by some pest insects. These are useful for monitoring the adult forms of pest moths, certain beetles and weevils, and some species of flies and fruit flies (Figure 2-2). Certain other materials are also used as trap attractants. For example, ammonium carbonate attracts many different species of flies; foodlike odors attract certain insects or rodents.

For monitoring, pheromones and other attractants are used in sticky traps or rodent spring traps (Table 2-1). Inside a building where food is stored, you can use these attractant traps to locate sources of infestation.

The effectiveness of attractant traps inside buildings is influenced by the number of traps used and where they are placed. Figure 2-3 shows a typical way of placing pheromone traps for monitoring stored-product insects. In small areas, use one trap to each 250 to 500 square feet of storage space. Larger areas such as warehouses require one trap to every 1000 to 2000 square

TABLE 2-1

Pheromones and Other Commercially Available Attractants for Pests Inhabiting Buildings and Storage Areas.

Pheromones and attractants for the following pests are available from various suppliers:

almond moth
ambrosia beetle
angoumois grain moth
cigarette beetle
confused flour beetle
drugstore beetle
fruit flies
house fly
Indianmeal moth
khapra beetle
lesser grain borer
Mediterranean flour moth
raisin moth
red flour beetle
sawtoothed grain beetle
tobacco moth
warehouse beetle

FIGURE 2-2.

Pheromone traps are useful for monitoring the activity of certain pests in buildings.

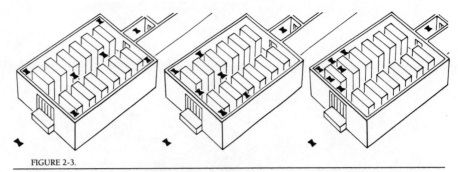

FIGURE 2-3.

One way of using pheromone traps in a building such as a warehouse is illustrated here. Begin by distributing traps throughout the building as shown by the drawing at the left. After several days, move traps that caught few or no target pests closer to the traps where catches were higher, as shown in the center drawing. Repeat this process again several days later, as shown at the right, to pinpoint the source of infestation. Keep a trap near each entrance at all times.

feet. Keep traps away from doors, windows, or bright lights, which may repel the target insects. Trap design (Figure 2-4) can affect the results of the trapping program—some styles appear to work better with certain species of insects. Check the supplier's recommendations for the most effective trap style. Also, choose a style based on the type of location where traps will be used; "wing" traps hung overhead, for instance, hide trapped insects from the public's view.

Check traps regularly. For insects, check traps once or twice per week at a minimum and remove all captured insects. Rodent traps must be checked every day. Clean or replace sticky surfaces whenever they become covered with debris. Replace pheromone lures periodically because they lose their effectiveness over time. Consult the manufacturer's guidelines for the effective duration of the attractant you are using.

Record the number of target insects removed from traps each time they are checked. Plot trap catches on a per-day basis, using a simple graph like the one illustrated in Figure 2-5. This will allow you to perceive changes in the insect's activity and verify the success of control measures. Compare this activity with activity in traps in other locations.

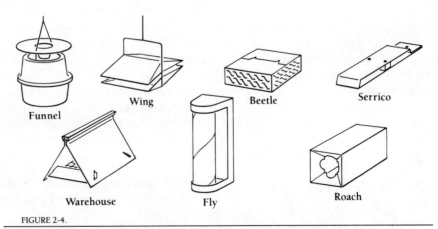

FIGURE 2-4.

Several styles of pheromone traps are available, depending on the type of pest and type of location being monitored.

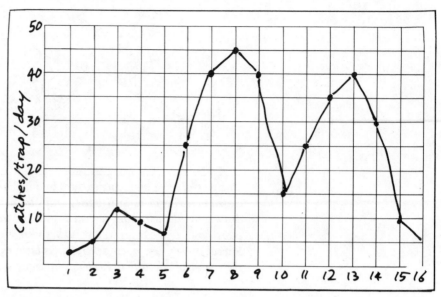

FIGURE 2-5.

A simple graph like this can help visualize the periods of peak activity of insects being monitored with pheromone traps.

FIGURE 2-6.

Traps equipped with ultraviolet or "black" lights attract some species of flying insects. These traps are most effective in enclosed areas, usually at night.

Light Traps. Traps equipped with ultraviolet lights, or black lights, attract several species of flying insects (Figure 2-6). These traps usually have a container with a funnel-shaped entrance that allows insects to enter easily but blocks their escape. Some light traps have an electrically charged grid that kills insects as they approach the light. Electrocutor traps are not used for insect monitoring.

Use light traps inside warehouses, grocery stores, and other enclosed areas for monitoring adult stages of flies, some species of fruit flies, and some stored-product insects such as Indianmeal moths and almond moths. Light traps are not effective for monitoring insects outdoors.

In a large building, use one blacklight trap for every 1000 square feet of floor space. Locate traps so the light is visible from all directions, but avoid placing them near windows or doors where the light may attract insects from outside. For monitoring stored-product insects, put traps in areas of the building where pest insects are most likely to be found—usually near a food source—but keep traps at least 5 feet away from food preparation or processing areas. Keep traps low if you are attempting to attract day-flying insects such as houseflies; in this case, traps should be no more than 5 feet above the floor.

Clean out blacklight traps at least once a week to prevent dead insects from becoming food for carpet beetles or other dermestids. Record the number and identity of the insects removed from the traps. Use this information to determine locations of greatest infestation and to detect cycles of pest outbreak. This information will also help you evaluate control efforts.

Blacklights are less effective in bright sunlight or where sodium vapor lights are being used. Also, the ultraviolet tubes used in light traps gradually lose their attractiveness to insects over time. Tubes should be replaced once a year.

Flypaper. You can use flypaper for monitoring flies within confined areas. Some manufacturers add a fly pheromone to the sticky coating to make it more effective.

Space several flypaper traps evenly throughout areas being monitored to find out where flies are concentrated. Do not use flypaper in dusty areas because

FIGURE 2-7.

Glue boards can be used to monitor or control insects and small rodents. They lose their effectiveness if the sticky substance becomes coated with dust or debris.

FIGURE 2-8.

Tracking powders are used to monitor the activity of small rodents and insects. Toxicants sometimes are added to the powder to poison the animal.

accumulated dust will clog the sticky surface and prevent flies from being caught. Flypaper traps are unsightly, so locate them out of the public's view. Check and replace traps frequently. Examine the captured insects to determine their identity. Keep records of the numbers and species of flies that were caught and use this information for selecting and evaluating control methods.

Glue Boards and Traps. Glue boards are occasionally useful for monitoring crawling insects, mice, and, in some instances, rats (Figure 2-7). Glue boards are sometimes used to locate areas where cockroaches congregate. By examining individuals caught on the sticky surface, you should be able to identify the species and perhaps determine areas of heavy infestations. Glue boards may also enable you to identify other types of insect pests within a confined area. Glue boards become ineffective when they are coated with dust, debris, or moisture.

For monitoring cockroaches, place glue boards along travelways next to intersections of walls and floors.

To monitor rodent activity, set glue boards along known runways and near areas believed to be nesting sites. Be sure to check the traps daily and dispose of any in which a rodent has been caught.

Mice and rats can also be monitored with spring-type or multiple catch traps. Place traps along runways and near nesting areas. Check these daily and remove captured rodents. Traps are most effective when they are baited with a substance that attracts rodents. See the "Trapping" section on page 189 for methods of baiting rodent traps.

Nontoxic Tracking Powder. Nontoxic tracking powders are fine dusts that provide a visual record of rodent or insect activity. Nontoxic tracking powder offers an alternative to glue boards and spring traps, which must be checked daily. These powders are also a safe way to evaluate the success of a control program.

Talcum or baby powder can be used for tracking pests, and commercial tracking powders are available that fluoresce under ultraviolet light, making it easier to locate pest trails. Avoid using any powders in areas where they might contaminate food.

Spread a thin layer of tracking powder evenly over surfaces where pests are known or suspected to occur (Figure 2-8). For easy cleanup when monitoring is completed, spread the tracking powder on sheets of paper. Look for tracks left in the powder as evidence of pest activity. Sometimes a trail of the powder leads to nests or hiding places.

ESTABLISHING THRESHOLDS FOR ACTION

Pest control decisions are influenced by health or safety dangers created by the pest, by legal restrictions on pest infestation, and by levels of pest tolerance. Occasionally a pest control decision depends on the costs involved to control a pest weighed against the benefits received. On the basis of any of these factors, a threshold for action can usually be established to determine what type of control is needed and when control should begin.

Health and Safety Threshold

Health or safety threats commonly require fast, extensive, and sometimes costly pest control measures. Several pests associated with structures, stored food products, food preparation facilities, hospitals, and other areas have the potential for causing injuries to people (mosquitoes, biting bugs, fleas, spiders, bees, and wasps, for example) or transmitting diseases to people or animals (rats and mice, cockroaches, fleas, flies, and mosquitoes). Some others, such as rodents, fungi, termites, and wood-boring beetles, cause the type of damage that makes structures unsafe or reduces their value.

Decisions to control pests are based on knowledge of the potential harm they can cause. If serious injury or damage may result, the control threshold must be very low. For instance, one rat chewing on electrical wiring can cause a serious fire.

Legal Thresholds

Health codes, marketing orders, and other regulations set limits on the amount of pest damage or contamination allowed in food products offered for sale or transported to other areas. Public safety codes often require control of pests in public buildings, commercial housing, food service facilities, and other public structures. Building and safety standards address the control of structural pests as well as the repair of damage caused by them. These legal thresholds dictate when pest control methods must be used, even though in some cases control methods cannot be economically justified or the pests may not be causing a hazard to public health or safety.

For information on laws that regulate pest infestation in certain buildings and on foods, contact state and local health departments and housing and community development offices. The Structural Pest Control Board has information on laws that apply to the control of structural pests. Federal marketing orders list allowable tolerances of specific pests or pest damage in fresh and stored food items; this information can be obtained from the United States Department of Agriculture, Agricultural Marketing Service.

Pest Acceptance Threshold

People have different degrees of acceptance of pests that they are willing to tolerate in their homes and workplaces. Pest acceptance thresholds may be high because of social or cultural factors or because of concerns about the costs or hazards of pest control methods used. A pest acceptance threshold can be extremely low due to a person's revulsion or fear of the pest. Acceptance thresholds may sometimes be modified if you can provide factual information about specific pests, the potential for pest damage, and methods of pest control.

Economic Threshold

In certain instances, the cost of control measures may need to be justified. Economic thresholds may apply if there are no health and safety, legal, or tolerance thresholds that need to be considered. An economic threshold is a level of pest abundance at which the potential loss caused by pest damage is expected to be greater than the cost of controlling the pest.

PEST CONTROL METHODS

Pests can be prevented, through sanitation and habitat modification, or they can be controlled by trapping, pesticide use, and, in some instances, biological control. Pests in structures are usually more effectively controlled when a combination of compatible control methods can be used.

Sanitation and Habitat Modification

Habitats are areas within the larger environment that are suitable for a pest's survival. Habitats provide a pest with some or all of its necessary living requirements such as food, water, shelter, optimum temperatures and humidity, and protection from enemies. A habitat can only accommodate a maximum number of pests due to limitations of one or more of these requirements. This maximum number is known as the *carrying capacity*. Where large quantities of food are available and shelter and other requirements are ample, the carrying capacity is high. Such a habitat can support an almost unlimited number of individuals of a pest species. If the carrying capacity is limited, however, the population tends to remain fixed in size. If you remove individuals from a habitat through pest control measures or if they die off due to natural causes, these individuals will be replaced by others, usually soon, unless the carrying capacity is reduced at the same time. Population size is maintained at the carrying capacity by increased reproduction among remaining individuals or by new individuals migrating in (Figure 2-9). Table 2-2 lists ways that modifying a habitat lowers its carrying capacity.

Habitat modification usually involves improving sanitation practices. Sanitation includes removing food, water, breeding sites, and shelter used by pests (Figure 2-10). Outdoors, you may need to trim or remove dense, pest-harboring vegetation near buildings, clean up trash, keep garbage in closed containers, provide for drainage of standing water, clean up animal wastes and spilled animal feed, and eliminate items that attract pests. Inside, sanitation includes storing foods and food wastes in tightly closed containers, cleaning up spills and residues, removing trash and other materials that can be used for nests, and thorough vacuuming and dusting on a regular basis. The

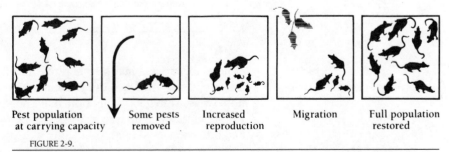

| Pest population at carrying capacity | Some pests removed | Increased reproduction | Migration | Full population restored |

FIGURE 2-9.

Populations of pests tend to remain fixed in size due to the carrying capacity of an area. In a building, for example, the carrying capacity is restricted by space, food, and water. If individuals are removed by some pest control method, the population may soon return to its original size as remaining individuals increase reproduction or as new individuals migrate into the area.

FIGURE 2-10.

Sanitation practices are helpful in reducing pest problems because good sanitation restricts the pest's access to food, water, or shelter.

TABLE 2-2

How to Modify Habitat to Lower the Carrying Capacity for Certain Pests.

MODIFICATION	EFFECT ON PESTS
Caulk cracks in cabinets, moldings, and other areas.	Reduces amount of hiding and nesting places available for cockroaches and certain fabric pests. Excludes ants and other crawling insects.
Seal food in pestproof containers or place in refrigerator or freezer.	Eliminates access to sources of food for cockroaches, ants, stored-product pests, flour beetles, rats, and mice.
Increase light and ventilation.	Makes area unsuitable for fungi and molds, subterranean termites, flea larvae, and other insects such as cockroaches and silverfish by reducing moisture. Reduces condensation, which can be a water source for many different pests.
Reglue loose wallpaper, patch peeling plaster or paint.	Reduces hiding and nesting sites for cockroaches and silverfish.
Remodel to eliminate false bottoms in cabinets and other structural voids or blow sorptive powder into wall voids, beneath cabinets, and in other inaccessible areas.	Eliminates hiding places for cockroaches, rats, mice, spiders, silverfish, firebrats, and other pests.
Insulate water pipes (seal well to prevent insulation from being a habitat for cockroaches).	Prevents condensation, which provides free water for some pests.
Store food wastes in tightly closed containers; remove from building frquently and clean containers.	Eliminates food sources for fruit flies, cockroaches, flies, mice, and rats.
Clean up food or liquid spills quickly.	Prevents attracting flies, fruit flies, cockroaches, rats, and mice.
Thoroughly clean food preparation and eating areas at least daily.	Eliminates food sources for fruit flies, cockroaches, flies, rodents, and other pests.
Launder or dry clean clothing and linens before storage; seal in heavy plastic.	Reduces problems with fabric pests.
Blow sorptive powder into wall voids, beneath cabinets, and in other inaccessible areas.	Denies hiding places for cockroaches due to its repellency.

cleaning of surfaces may also improve the effectiveness of pesticides by removing grease, oils, dust, and other contaminants that interfere with their function. To assist in good sanitation, make sure interior areas are well lighted to simplify cleaning and easy detection of pests and pest damage. Sweepings and other wastes should be taken to a disposal area outside of the building.

Other sanitation practices include removing dirt mounds, wood pieces, and other cellulose debris from areas beneath buildings to keep from promoting termite or rodent problems. To protect against fungi, never leave unprotected wood in contact with soil or other sources of moisture such as leaking pipes or faulty drains. Provide adequate ventilation to areas beneath buildings to reduce moisture.

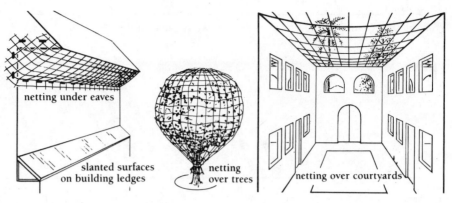

FIGURE 2-11.

Various methods can be used to keep birds from roosting or building nests on buildings. Netting is helpful in excluding birds from trees or courtyards.

Outdoor lights placed near entrances to buildings attract many flying and crawling insects at night. If possible, locate light fixtures away from entrances. Otherwise, modify the type of light being used. Sodium vapor lights are better than mercury vapor lights or standard incandescent lights for outdoor use because they emit a spectrum of light that is less attractive to insects; yellow "bug" bulbs work on the same principle.

Birds can become pests when they use outside surfaces of buildings for roosting or nesting. Birds generally prefer flat surfaces that offer protection from wind, rain, and extreme temperatures. Cliff swallows build their mud nests under overhanging ledges or roofs for the same reason. To prevent birds from roosting or nesting, use plastic or wire mesh screening, cloth netting, or metal flashing. You can also modify flat surfaces or overhangs to make them unsuitable to birds (Figure 2-11). The ultimate decision on how best to deal with pest birds usually depends on the species that are causing problems, their location on the building, and physical features of the building.

A program of sanitation and habitat modification requires cooperation between pest control specialists, building owners and inhabitants, housekeeping staff, and building and landscape maintenance workers. It is necessary that everyone understand how these practices influence pest problems (Table 2-3, next page). People living or working in a building must keep food, food waste, and trash in pestproof containers and store other items in designated places where they cannot attract pests. Inhabitants should promptly report pest problems. Housekeeping and landscape maintenance workers can help by keeping interior and exterior areas free of trash, nesting sites, and other items that might be attractive to pests; they should provide containers for wastes and specify locations for storage of other materials. Buildings must be monitored on a regular basis to ensure that sanitation conditions are maintained and to spot new problem areas as they occur. Tenants and persons responsible for housekeeping and landscape maintenance must be notified of conditions that promote pest buildup so they can take corrective action.

FIGURE 2-12.

Pests may be excluded from a building through good construction techniques or by adding weatherstripping, caulking, or other materials as shown in this illustration.

Exclusion. Exclusion is a type of habitat modification useful for keeping fleas, ants, cockroaches, stored-product pests, termites, rodents, and other pests from entering buildings (Figure 2-12). The design and construction of a building may either promote pests or exclude them. Pestproof design and construction should be an important consideration when planning new structures and remodeling older ones.

TABLE 2-3

Factors that Contribute to Pest Problems In and Around Buildings.

	Shelter	Animal Wastes	Debris or Clutter	Firewood	Garbage	Flours/Grains/Cereals	Cardboard/Newspapers	Vines/Shrubbery	Soiled Clothing	Spiderwebs	Spoiled Fruits/Vegetables	Stores Linens	Sugar/Sweets	Moisture or Free Water	Weeds	Sawdust/Wood Scraps	Humans, Dogs, Cats, Other Pets	Structural Lumber	Furniture/Cabinets
Ants		■			■			■			■		■	■					
Bats	■																		
Bees					■						■		■	■					
Birds	■					■		■						■	■				
Biting bugs	■		■																
Carpenter ants	■			■									■	■		■		■	
Carpenter bees	■			■															
Carpet beetles							■		■	■		■							■
Clothes moths									■			■							
Cockroaches	■	■			■	■	■	■			■		■	■					■
Decay fungi														■					
Firebrats	■		■			■	■				■			■					■
Fleas	■														■		■		
Flies		■			■						■		■				■		
Flour beetles						■													
Fruit flies					■						■		■						
Grain moths						■													
Granary weevils						■													
Mice	■	■		■	■	■	■	■			■		■	■	■				
Mites		■	■			■											■		
Rats	■	■		■	■	■	■	■			■		■	■	■		■		
Silverfish	■		■				■	■				■		■					■
Spiders	■			■			■	■							■				■
Termites	■	■		■				■					■			■		■	
Ticks								■							■		■		
Wasps	■	■			■			■			■		■	■					
Wood-boring beetles				■										■		■		■	■

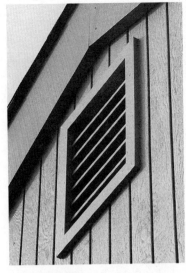

FIGURE 2-13.

Properly screened vents prevent rodents and other pests from gaining access.

Check building exteriors for ways that insects, rodents, or other pests can enter. Obvious entrances for many types of pests are doorways and windows. These must be fitted with tight-fitting screens or doors. Properly installed weatherstripping eliminates small cracks that provide access for some pests. Also, check attic and foundation vents to ensure that they are tight and screened to exclude rodents (Figure 2-13). Look for foundation or wall cracks, gaps in siding or joints, and areas where pipes, wires, or other objects pass through walls. Fill openings with concrete or another suitable patching material, or cover openings with metal flashing. Inspect chimneys and roof vent pipes for adequate screening.

Select pest exclusion materials according to the type of pests encountered. Refer to Table 2-4 for selecting materials used for excluding pests and repairing openings. Insects can gain entry through extremely small openings. Mice are able to squeeze through cracks as small as ¼ inch, and many rats manage to get through ½-inch openings. Rodents can chew through wood, thin metal, caulking, soft patching compounds, and even some concrete mixtures. Bats, on the other hand, do not chew through walls, roofs, or other surfaces, but enter structures through existing openings ⅜ inch or larger. Exclude bats by using thin wire mesh, sturdy cloth mesh, or almost any type of well-secured patching material.

TABLE 2-4

Materials Used for Excluding Pests and Repairing Openings in Structures.

MATERIAL	SPECIFICATIONS	USES	PESTS EXCLUDED
Bird netting	¼-inch mesh plastic or cloth	Under eaves, around roof openings	Birds, bats
Brick	Must have strong mortar seams	Protective barrier for structural wood	Rats, mice, most insects
Caulking	Architectural grade, must be flexible; silicone type works best	Fill cracks and small holes in wood, masonry, and plaster	Ants, cockroaches, bats, spiders, carpenter ants, wasps
Concrete	1:2:4 mixture (cement:sand: aggregate—use ⅜ inch or smaller aggregate); add water to a wet sand consistency	Patch holes in walls or construct barriers	Rats, mice, insects
Door sweeps	Metal, leaving less than ¼ inch gap at bottom of door	Close gaps at bottom of doors	Rats and mice, bats
Duct tape	Heavy duty	Temporary seal for large cracks, holes, seams	Bats, some insects
Expanded metal	Heavy gauge galvanized metal or aluminum, mesh less than ¼ inch	Cover vents, large openings	Rats, mice, bats
Glass jars	Must have screw-on metal lids or tight-fitting plastic lids; jars with rubber seals and snap caps work best	Store small quantities of dried foods, sugar, syrup, honey	Stored-food pests, ants, wasps (outdoors), mice, rats
Hardware cloth	¼-inch mesh or smaller, 19 gauge galvanized metal	Ventilators, louvers, large openings, vents	Mice, rats, bats, birds
Insulation	Roll or blow-in type	Attics, wall voids	Bats

MATERIAL	SPECIFICATIONS	USES	PESTS EXCLUDED
Metal flashing or sheeting	19 gauge galvanized metal or 22 gauge aluminum	Roof valleys, construction joints, openings and holes in walls or roofs, or covering of exposed wood surfaces	Mice, rats, bats, termites
Metal grills	Heavy gauge galvanized metal or aluminum, slots should be ¼ inch or smaller in width	Cover vents, large openings	Rats, mice, bats
Metal threshold	Must seal with bottom of door	Closes gaps at door bottom	Rats, mice, ants, cockroaches, spiders, most other crawling insects
Mortar	1:3 mixture	Fill cracks in masonry and concrete	Rats, mice, crawling insects
Perforated metal	Heavy gauge galvanized metal or aluminum openings should be ¼ inch in width or smaller	Cover vents, large openings	Rats, mice, bats
Plastic bags	Heavy duty type (must be sealed well)	Store linens, woolens, cereals, sugar, flour, other dried foods and sugary foods	Clothes moths, carpet beetles, ants, cockroaches
Plastic containers	Must have tight-fitting lids	Store quantities of dried foods, nuts, grains, sugar, syrup, honey	Stored-food pests, ants
Putty	Nonshrinking, weatherproof; silicone type works best	Fill cracks and small holes in wood, masonry, and plaster	Ants, cockroaches, spiders, wasps
Self-expanding polyurethane foam		Fill large voids and irregular openings, seams in corrugated tile and metal roofing	Bats, ants, cockroaches, spiders
Silicone rubber	Caulking type	Fill cracks and small holes in cabinets, baseboards, moldings, around windows, tubs, other areas	Ants, cockroaches, spiders, wasps
Steel wool	Fine grade (#00) tightly packed into hole, seal with caulking	Plug holes in wood (rusts when exposed to moisture)	Carpenter ants, carpenter bees (can be used to temporarily plug holes to exclude mice)
Weatherstripping	Rubber or felt	Seal cracks around doors and windows	Bats, ants, cockroaches, spiders, other small insects
Window screening	Galvanized metal or aluminum	Vents, windows, doors	Bats, birds, spiders, flies, bees, wasps, mosquitoes, other flying or crawling insects

Pests can also be brought into buildings on items such as those listed in Table 2-5.

Inspection. Inspect items brought into a building for pest infestation. For example, firewood may harbor carpenter ants, spiders, cockroaches, wood-boring beetles, termites, or similar pests, or eggs of some pests. Furniture, rugs, and other items moved from an infested building can be contaminated with cockroaches, carpet beetles, or fleas. Dogs and cats bring in fleas and ticks.

Managers of grocery stores, cafeterias, restaurants, or other food-handling establishments should work with pest control specialists in developing systems

TABLE 2-5

Ways Some Pests Gain Entry into Buildings. Pests may gain entry by being carried in on items such as those listed here.

ITEM	PESTS
Appliances	Mice, cockroaches
Books/papers	Cockroaches, silverfish, firebrats
Cardboard containers	Cockroaches, silverfish, firebrats, stored-product moths, spiders, mice, rats (occasionally)
Carpets/rugs	Carpet beetles, fleas, cockroaches, clothes moths
Clothing	Clothes moths, lice, fleas, carpet beetles
Cut flowers	Carpet beetles, spiders
Firewood	Spiders, wood-boring beetles, termites, carpenter ants, cockroaches
Fruits/vegetables	Fruit flies, spiders, ants
Furniture	Spiders, wood-boring beetles, cockroaches, fabric pests, fleas, sometimes mice
Grains/cereals	Stored-product beetles and moths, cockroaches, mice in bulk containers
Groceries/dry goods	Cockroaches, spiders, silverfish, firebrats, mice, rats (occasionally)
Lights near entrances	Spiders, carpet beetles, flying insects
Pets	Fleas, ticks
Plants	Ants, spiders, mites
Vacuum cleaner bags	Fleas, cockroaches, carpet beetles, fabric pests

for examining bulk containers for cockroaches and stored-product insects. Small packages of certain items suspected of being infested can be placed in a freezer for a few days to destroy insects. Persons responsible for purchasing can help by buying from suppliers that can deliver pest-free merchandise. Some manufacturers are now using pheromone traps in shipping containers as a way of monitoring the pest-free status of their products (Figure 2-14).

Trapping

Besides their benefits as monitoring devices, traps are used to kill pests or to catch pests so they can be removed from an area. Many types of vertebrate and invertebrate pests can be controlled through trapping. Traps do not require the use of potentially hazardous chemicals, and the user can easily view the success of the trapping program. However, successful trapping programs require skill, time, and attention to develop workable techniques. Even so, trapping may not always work well enough under some conditions to satisfactorily control target pests. Trapping techniques that are successful

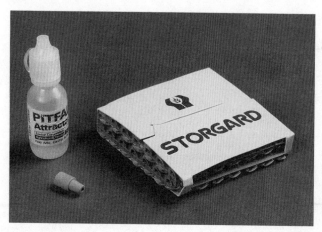

FIGURE 2-14.

A special type of pheromone trap may be placed in shipping containers to detect contamination during transit and to ensure that the product is pest free.

in one situation may not always work as well under different conditions or at other locations.

Traps include snap traps, glue boards, live animal traps, pheromone traps, and light traps. Specific uses of trapping devices are described in detail in other parts of this book. For example, the use of snap traps, glue boards, and live animal traps for control of rats and mice is discussed in Chapter 10.

Pesticides

The application of pesticides is the most common pest control method used in and around buildings, enclosed areas, and vehicles. Some pesticides provide chemical barriers to prevent insects from getting in. Pesticides are also used to treat soil, wood, fabrics, and other items to prevent pest damage.

Pesticides are available as baits, tracking powders, desiccants (inert dusts or sorptive powders), liquids, dusts, and gases. The type of pesticide used and the kind of formulation selected is based on the life habits of the pest, its density, and its location.

The following chapter discusses methods of safely using pesticides.

Biological Control

Biological control is gaining more importance as a pest control method for certain insects in structures. For instance, parasitic nematodes are occasionally effective in controlling some species of termites. Cockroach populations have been successfully reduced in certain locations by introducing parasitic wasps. Biological control techniques either augment other control practices or replace more disruptive or hazardous methods.

Biological Control

Biological control of pests involves the use of one living organism to control another. For example, most arthropod pests have natural enemies or disease organisms that control or suppress them effectively under some conditions or in some situations. Occasionally insects or microorganisms contribute to the control of certain weeds. Microorganisms also provide a degree of natural control of pest birds and rodents. Sometimes biological control can be an important component of a pest management program by taking advantage of these helpful organisms.

When natural enemies or microorganisms are present, efforts can be made to protect them so they may increase in number and help control pests more effectively. Large numbers of the beneficial organism may be introduced into an area to control a pest. Certain commercial companies specialize in producing beneficial organisms for pest control programs.

3 Using Pesticides Safely

When you apply any pesticide, you assume the legal responsibility for using it strictly in accordance with label instructions. You must always protect people who live or work in the treated area so they are not exposed to harmful residues. Avoid using pesticides or application methods that might injure nontarget animals or plants or damage property. Pesticide use should not endanger the environment or cause contamination of groundwater, soils, air, or human and animal foods. In addition, you must use pesticides in ways that avoid excessive exposure to any part of your own body.

This chapter gives a brief introduction to pesticide types, formulations, hazards, and safety precautions for residential, industrial, and institutional pesticide uses. More detailed information can be obtained in Volume 1, *The Safe and Effective Use of Pesticides*. This information is summarized in the following five boxes. These provide precautions that must be observed when handling pesticide containers, guidelines for mixing pesticides, some of the steps that must be taken to properly apply pesticides, ways to safely store these materials, and information on pesticide disposal.

Handling Pesticides

Undiluted pesticides in their original containers must be handled carefully. Wear rubber gloves and protective clothing, such as a waterproof apron, when handling pesticides. Do not drop or throw containers or packages because this may cause damage and leaks. Check for contamination or leaks on all packages being handled, and do not let damaged packages or spilled pesticide come in contact with your skin or clothing. If a container is damaged and leaking, the pesticide should be transferred to another container and must be properly identified. When working around a leak, you may need to wear respiratory and eye protection—check the pesticide label for all precautions and required safety equipment. Never walk through a spilled pesticide.

Never leave pesticide containers unattended or stored in unlocked areas. Always keep pesticides away from food and water and away from sources of heat and fire. Never allow paper containers to get wet.

Do not eat, drink, or smoke while handling pesticides and pesticide containers. Wash thoroughly when you finish handling pesticide containers and before eating, drinking, smoking, or using the bathroom.

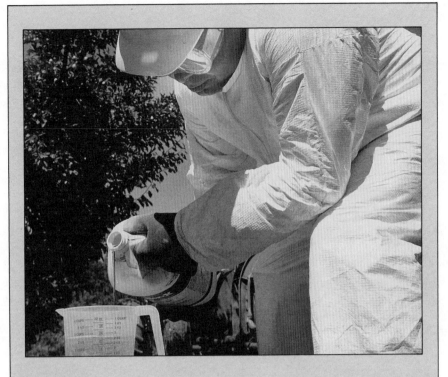

Mixing Pesticides

Pesticides must be properly mixed to ensure that the correct amount of pesticide is thoroughly incorporated into a measured amount of water or other solvent. Techniques for mixing pesticides are the same for large and small volumes. Before beginning, read the mixing directions on labels of all pesticides you will be using and decide on the proper order that chemicals should be added to the spray tank (see below). If adjuvants are needed, these are usually added before pesticides unless label instructions give a different order.

General rules for mixing pesticides include:

- Determine what protective clothing is required for mixing by checking the pesticide label.

- Before adding pesticide to the spray tank, check for leaks in the tank and hoses. Make sure the equipment is clean and operating properly.

- Use only clean water in the spray tank. Be sure the pH of the water is within a range suitable for the pesticide(s) being used. Buffers or acidifiers may be required to adjust the pH.

- Measure pesticides carefully, accurately, and safely to be certain the correct amount of pesticide is put into the spray tank.

- When mixing more than one pesticide into the spray tank, add the materials in the following order:

  1. Wettable powders
  2. Flowables and dry flowables
  3. Water-soluble concentrates
  4. Emulsifiable concentrates

Applying Pesticides

To apply pesticides correctly, you must make sure that the correct amount of active ingredient is applied to the area to be treated and that the pesticide is confined just to that area. Here are some important things that you must do to properly apply pesticides:

- Calibrate the application equipment accurately.

- Use the correct amount of active ingredient. Check the pesticide label for rates of application.

- Measure the area to be treated so that the correct quantity of pesticide mixture can be prepared.

- Check the application site for hazards that might affect the safety of the application. Hazards include electrical outlets and exposed wiring, sources of ignition such as flames or sparks, confined spaces, and improper ventilation. They also include irregular surfaces over which the applicator or equipment must travel.

- Make sure that the weather conditions are suitable for pesticide application. This includes temperature, humidity, and, if making outdoor applications, wind, fog, and rain.

- For liquid sprays, control the droplet size and spray pressure to prevent drift and to keep the spray on target.

- Set up an application pattern that prevents you from having to walk or drive through treated areas.

- Do not apply pesticides in or near air conditioning or heating vents or ducts.

- Keep people and animals away from the area during application and until the treated area is safe to reenter.

Storing Pesticides

Store pesticides in their original, tightly closed containers. Whenever possible, wipe or wash pesticide residue off the outside of containers before they are put into storage. Protect pesticides from extremes in temperature and from becoming wet. A pesticide storage area should be a separate building, away from people, living areas, food, animal feed, and animals. The area must be well ventilated, well lighted, dry, and secure, with lockable doors and windows. Post signs near all primary entrances to warn others that the building contains pesticides.

Some pesticides do not store well for long periods of time. Extended storage, especially after temperature extremes, may cause chemical changes resulting in some products losing their effectiveness or others becoming more toxic. Moisture and air picked up during storage may alter the composition of some pesticides, especially those stored in unsealed containers. Solvents and petroleum-based chemicals can degrade some types of containers after a period of time.

Most pesticide chemicals should not be stored for longer than 2 years. Before pesticides exceed their shelf-life, use them in an appropriate application or transport them to an approved disposal site.

Pesticide Disposal

Leftover pesticide mixtures are considered hazardous wastes unless they can legally be used to control pests in another site. Therefore, whenever possible, mix up only the amount that is required for each job. Excess pesticide must never be indiscriminately dumped; such dumping is a potential source of environmental and groundwater contamination and is illegal. Persons convicted of dumping are subject to large fines and jail terms.

Rinse water from cleaning of equipment is also a hazardous waste and must be treated accordingly.

Hazardous materials such as leftover pesticide residues must be transported to an approved Class 1 dump site, or they may be rendered nontoxic by means of a treatment with ultraviolet light and ozone.

Pesticide containers must be triple rinsed before they can be disposed of in a Class 2 disposal site.

Check with the California Department of Health Services, the Water Quality Control Board, and the local Agricultural Commissioner for methods of disposing of hazardous pesticide wastes and empty pesticide containers.

PESTICIDE TYPES

The most common pesticides used in and around structures are insecticides and rodenticides. Table 3-1 compares the toxicity categories of various insecticides used in structures, illustrating that some of the materials are more acutely hazardous than others. Occasionally fungicides or herbicides are used to control pests near buildings. Wood preservatives are a special class of pesticide used to protect structural and decorative wood, utility poles, and marine pilings.

Pesticides used for residential, industrial, or institutional pest control are available in several types of formulations (Table 3-2). The following box

TABLE 3-1

Common Materials Used to Control Insects in Structures. Insecticides used for the control of insect pests in structures may belong to any of the three toxicity categories. (Note: Some materials on this list may no longer be registered as insecticides in California, or the registration may no longer include structural uses. Before using any pesticide, check the current label to be certain it is registered for the intended use.)

COMMON NAME	PRODUCT NAME	CATEGORY
acephate	Orthene	III
	Whitmire PT 280	III
bendiocarb	Ficam "D"	III
	Ficam "W"	II
	Ficam Plus	II
benzeneacetate	Pyrid	II
boric acid	Perma-Dust (PT 240)	I
carbaryl	Sevin	III
chlorpyrifos	Dursban LO	II
	Dursban 2E	II
	Dursban TC	II
	Killmaster II	II
cypermethrin	Demon WP	II
	Demon TC	II
	Demon EC	II
diazinon	Diazinon 2D, 4E, and 4S	II
	Knox-Out	II
	Whitmire PT 260	II
	Diazinon 3 Dust	II
dichlorvos	Vapona Aerosol	I
hydramethylnon	Combat Roach Control	II
hydroprene	Gencor	III
methroprene	Precor	III
methylmercury	Tribute	II
permethrin	Torpedo	III
proptamphos	Safrotin	III

TABLE 3-2

Formulation Types and Their Uses.

TYPE	USES	COMMENTS
Wettable powder	Insecticides, fungicides, herbicides.	Require agitation during application. May leave visible residues on treated surfaces after drying. Do not penetrate surfaces well.
Dry flowable	Insecticides, fungicides, herbicides.	Require agitation during application. May leave visible residues on treated surfaces after drying. Do not penetrate surfaces well.
Soluble powder	Insecticides, fungicides, herbicides.	Penetrate better than wettable powder or dry flowable formulations.
Emulsifiable concentrate	Insecticides, fungicides, herbicides.	May damage plants. May cause spotting or staining of treated surfaces.
Flowable	Insecticides, fungicides, herbicides. Limited number of pesticides available in this formulation.	Require agitation during application. May leave visible residues on treated surfaces after drying. Do not penetrate surfaces well.
Water-soluble concentrate	Insecticides, fungicides, herbicides, and rodenticides (as drinking solutions). Limited number of pesticides available in this formulation.	Penetrate better than wettable powder or dry flowable formulations.
Low-concentrate solution	Ready-to-use insecticides.	High cost per unit of active ingredient. Very useful in control of insects in buildings. Usually do not cause stains.
Fumigant	Insecticides, sometimes used for control of rodents.	Must be used in tightly closed area. Require special application techniques and equipment.
Dust	Insecticides, rodenticides.	Highly visible residues. Must be kept dry. Some types may be used as tracking powder.
Granule/pellet	Rodenticides.	Formulated as baits to be used in bait stations.
Microencapsulated formulation	Insecticides.	Used for slow, sustained release of active ingredient over time.
Bait	Insecticides, rodenticides, avicides.	May be attractive food substance coated on or impregnated with toxic material. Used in bait stations or bait blocks or scattered in safe location.
Impregnate	Insecticides, fungicides.	Include items such as flea collars, pest strips, and special paints. May also include factory-treated fabrics or carpets.

defines some of the common formulations and describes their advantages and disadvantages. Insecticides and rodenticides can be applied as baits (treated grains, meals, or other substances), liquids, granules, gases, or dusts. Some insecticidal dusts are used as tracking powders or desiccants. Fungicides and most herbicides are generally applied as liquid sprays, although some herbicides are available in a granular formulation.

Liquids

Pesticide liquids are mixtures of powdered or liquid active ingredients combined with liquid carriers such as water or oil. Pesticides may dissolve in the

Pesticide Formulation Types

Wettable Powder (W or WP). The pesticide is not soluble in water. It is combined with a finely ground material such as clay and combined with other ingredients to improve mixing. Most wettable powders contain between 15% and 75% active ingredient. When mixed with water, they form a suspension that must be kept agitated during application. Wettable powders may be abrasive to nozzles and pumps, but they are one of the safest formulations for use on plants for insect or disease control. Wettable powders usually leave a visible residue.

Dry Flowables or Water-Dispersible Granules (DF or WDG). This formulation is similar to a wettable powder but is in the form of granules that must be mixed with water before use. These pesticides require agitation during use. Usually there is a higher percentage of active ingredient in this formulation than in wettable powders. These are measured out by volume rather than weight.

Soluble Powders (S or SP). Soluble powders dissolve with water in the spray tank after mixing to form a true solution. These do not require agitation once they have been thoroughly mixed, and they are not abrasive to nozzles or pumps. Only a few pesticide active ingredients are soluble in water.

Emulsifiable Concentrates (E or EC). Emulsifiable concentrates are pesticides that are soluble in an organic solvent but not in water. When the undiluted formulation is combined with water, a milky emulsion is formed that must be kept agitated during application. ECs are not abrasive to application equipment, but the solvent may contribute to the deterioration of rubber and plastic seals and hoses of the application equipment. The solvent may also injure plant foliage and may cause damage to certain types of surfaces.

Flowables (F). Flowables consist of finely ground pesticide combined with a liquid solvent and emulsifiers. When mixed with water, they form a suspension similar to a wettable powder. Flowable formulations must be agitated during application. These may be abrasive to application equipment components. They leave visible residues on treated surfaces.

Water-Soluble Concentrates or Solutions (S). Water-soluble concentrates or solutions dissolve in water to form a true solution. Once dissolved, they do not require agitation. They are nonabrasive. Only a few pesticide active ingredients are soluble in water.

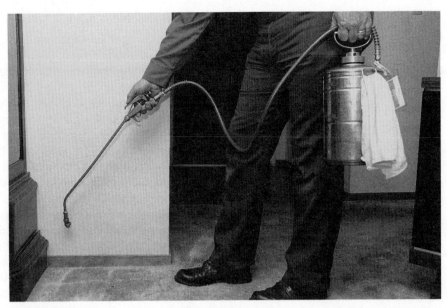

FIGURE 3-1.

Liquid pesticides can be applied with aerosol dispensers, hand-held compressed air sprayers, backpack sprayers, or larger, motorized spray units.

carrier to form a solution or may remain suspended in the liquid to form an emulsion or suspension. Suspensions and emulsions require constant agitation to maintain a uniform spray mixture.

Liquid pesticides are applied as spot treatments, crack and crevice treatments, fogs or mists in confined areas, or general sprays to large areas. The common ways to apply liquid sprays are with aerosol dispensers, hand-held compressed air sprayers, backpack sprayers, or larger, motorized spray units (Figure 3-1).

When liquid sprays are applied, a residue of pesticide active ingredient remains on the treated surfaces and helps to control pests over a period of time. The length of time depends on the type of pesticide used, the type of formulation, the concentration of active ingredient applied, the type of surface treated, and environmental influences such as temperature, humidity, or sunlight.

Undiluted pesticides contain concentrated amounts of active ingredient that may cause serious injury if inhaled, splashed or blown into the eyes, or spilled on the skin or clothing. Some concentrated pesticides may be flammable.

Applying liquid sprays in certain areas may be extremely hazardous. For example, electric outlets, motors, or exposed wiring pose a potential threat of electrical shock to persons applying water-based pesticide sprays. Pilot lights and gas flames from heaters and appliances may ignite flammable petroleum-based pesticides. Sparks from electric motors and switches and glowing heating elements may also ignite flammable materials (Figure 3-2). Pesticide vapors or fumes in confined areas may injure people if ventilation is inadequate.

FIGURE 3-2.

Many hazards may be found in an area being treated with liquid pesticides. Avoid application near electrical outlets, switches, motors, or heating elements.

Gases

Gases that kill pests are known as fumigants. Fumigants are used to control certain stored-product insects, drywood termites and wood-destroying beetles, soil-infesting nematodes, soil pathogens, and some rodents. The process of applying fumigants, or fumigation, is much different from other forms of pesticide application and requires special training and equipment.

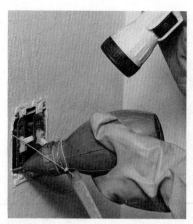

FIGURE 3-3.

Bulb applicators can be used to apply dusts to small, confined areas such as electrical outlets and cracks and crevices.

Dusts

Dust formulations are finely ground dry powders that contain toxic materials. These are sometimes used to control rodents and certain insects. Most dusts are blown into inaccessible places where pests hide. Dusts do not penetrate surfaces, and they usually break down slowly. Therefore, the active ingredient in dust formulations remains on the treated surface and is active against pests for a long period of time if the treated area stays dry. Because they do not penetrate, dusts are more effective than liquids on absorptive surfaces such as concrete.

Dusts may be applied in cracks and crevices, under cabinets or appliances, and in other areas inaccessible to children and pets. This formulation leaves visible residues on treated surfaces, which often limits its use to areas such as warehouses, attics, crawl spaces, and wall voids.

Dusts usually provide better coverage than sprays in inaccessible or hard-to-reach places. In wall voids, they can be dispersed with compressed air to reach all surfaces. During manufacture, dusts are sometimes given an electrical charge or they are combined with an electrically charged powder to make them cling to surfaces better. Bulb applicators (Figure 3-3), shaker cans, aerosol cans, and compressed air dusters are used to apply dust formulations.

When using dusts, prevent their drift into the airspace of living or work areas. Apply dusts only according to the instructions on the pesticide label. Always wear approved respiratory protection to avoid inhaling dust particles.

Toxic Tracking Powder. Toxic tracking powder is a dust formulation that may be useful where bait is not accepted by rodents or where there is an abundance of natural food. Target rodents pick up the toxic dust on their body surfaces as they walk through it and later ingest some during grooming. The toxic component of some tracking powders can also be absorbed through the animal's skin.

Apply toxic tracking powder to travelways alongside walls, inside wall voids, and in attics and crawlways. Blow the powder into inaccessible areas where rodents are known to travel. Avoid the use of power blowers in exposed areas to prevent dispersing the powder beyond the treatment site. Once control has been accomplished, remove remaining powder from parts of the treatment site that are exposed. To remove powder, use a vacuum equipped with an HEPA filter approved for pesticides. Once the area has been cleaned, dispose of the filter and vacuum bag in an approved hazardous waste disposal site.

Follow label directions for using a toxic tracking powder and carefully select locations where the powder is to be used. Do not put powder where it can be dispersed by air movement or tracked by pests onto food, eating utensils, or food preparation surfaces. Never use toxic tracking powder on shelves, cupboards, or ceiling beams overhead in food preparation or eating areas. Because of the hazards, do not use toxic tracking powder in food processing plants or food storage warehouses; never apply it in or around homes unless it can be applied inside wall voids or other inaccessible areas. In locations where people or animals may accidentally contact the powder, confine its use to bait stations; the combination of toxic tracking powder and toxic bait can sometimes be very effective.

Toxic tracking powder loses some of its effectiveness in damp areas because moisture causes the powder to cake and not stick to the animal's body; moisture may also speed the breakdown of the toxic material. Toxic tracking powder formulations are fast-acting poisons, therefore rodents die quickly. Rodents dying in inaccessible wall voids or other out-of-the-way areas could create odor problems or attract flies.

DESICCANT BEING APPLIED . . .

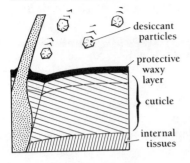

desiccant particles

protective waxy layer

cuticle

internal tissues

AFTER APPLICATION . . .

moisture escaping

wax adsorbed by desiccant particle

FIGURE 3-4.

Desiccants destroy insects and mites by removing or disrupting the protective outer body covering, as illustrated here. This causes the organism to lose body fluids.

Desiccants. Desiccants are dusts or sorptive powders used to control some insect pests found in buildings. The powder abrades or adsorbs the waxy coating that protects insects from losing water (Figure 3-4). Desiccants often last longer than other types of insecticides; however, insects must move through the dust and pick some up on their bodies for it to be effective. To apply, blow desiccants into wall voids, attics, and crawl spaces and dust them into other areas where insects hide. Some desiccants are highly repellent, which helps exclude insects from treated areas. Avoid breathing dusts during application by wearing respiratory protection.

Granules

Granular formulations are sometimes used to control ants, sowbugs, earwigs, snails and slugs, and occasionally other soil-inhabiting organisms. Usually granules are combined with a food substance or attractant to encourage target pests to feed on them. *Do not apply granules in areas where children or pets may find them.*

Certain herbicides are available in granular formulations. These can be spread evenly over the soil surface with a mechanical spreading device. Once the granules have been applied, the herbicide is generally incorporated into the soil by irrigation or cultivation.

Poisoned Bait

Poisoned bait may be used to control specific types of insects as well as snails, slugs, and rodents. Some birds may be controlled with poisoned bait if you first obtain a special permit from the California Department of Fish and Game.

Most baits are a combination of pesticide and food material. They may be in the form of powders, grains, granules, kibbles, or blocks. Baits are usually placed in a bait station or secured in protected places (Figure 3-5). Baits used to control snails or slugs, earwigs, or oriental cockroaches are usually broadcast over the soil around the outside of a structure.

FIGURE 3-5.

Poisoned baits are usually placed in bait stations to prevent children or nontarget animals from being exposed to the toxic material.

Table 3-3 is a guide to selecting bait types. Choose bait types and bait station styles on the basis of (1) the type of pest being controlled, (2) the past history of bait use, and (3) the conditions where baiting will take place. For example, when baiting for ants, select a bait that foraging workers will carry back to the nest to feed to the colony's reproductives and brood; the toxic substance must be slow-acting so that foraging workers are not killed before they reach the nest. Bait used to control flies, on the other hand, must be fast-acting to stop continued annoyance and prevent further egg laying.

Certain rodent baits contain an anticoagulant, which interferes with the animal's normal blood clotting process. It is important for rodents to feed on some anticoagulant baits over a period of several days so they will consume enough toxic material to be effective. If there is an interruption of feeding

TABLE 3-3

How to Select Bait Types. Follow these guidelines when selecting and applying baits.

PEST	TYPE OF ACTIVE INGREDIENT	HOW APPLIED	WHERE USED
Ants	Slow acting so workers can carry bait back to feed others.	Use bait stations such as metal ant stakes, hollow straws, or small containers.	Locate near nests, along trails, inside electrical boxes, and and around periphery of buildings.
Birds	Quick acting so poisoned birds will frighten away other birds, minimizing total bird kill. A permit may be required.	Use in bait stations or feeding troughs.	Hang in trees or near roosting sites.
Cockroaches	Quick acting for immediate knock down; slower acting for continued control.	Use in enclosed bait boxes with many small entrances.	Place under appliances, sinks, cabinets, and other concealed areas where cockroaches are found.
Flies	Quick acting so flies will not reproduce.	Apply as surface spray or use in feeding stations.	Apply sprays to resting surfaces. Locate feeding stations outdoors near garbage cans or other areas where flies congregate.
Rats and mice	Several types including acute toxics and single- and multiple-feeding anticoagulants. May also be added to drinking water. Vary types if bait shyness develops.	Use in bait stations or as bait blocks. Bait stations must be large enough to accommodate several individuals; should have at least two entrances. Use bait blocks in damp locations.	Locate near runways in protected areas. Choose several locations.
Wasps	Quick acting to reduce number of adults present.	Place in feeding stations.	Locate feeding stations around periphery of outdoor areas used by people.

FIGURE 3-6.

Bait blocks can be used without a bait station as long as they are located safely out of reach of children or nontarget animals. Paraffin blocks can be used in damp areas. The wax helps keep the bait fresh and prevents mold.

for longer than 48 hours, the animal will recover and accumulated toxic effects will be lost. To prevent this from happening, check and refill bait stations regularly. Other anticoagulant baits are effective after a single feeding.

Selecting a bait for rodent or bird control also depends on where it is to be placed. Toxic powders, poisoned grains, and granular formulations used indoors usually need to be confined to bait stations. Bait blocks can be used without a bait station, but place them where they can be secured and are out of the reach of children, pets, and nontarget animals. In damp areas, use rodent bait in the form of paraffin blocks that can withstand moisture (Figure 3-6); the wax keeps the bait fresh and helps prevent mold. Do not apply powdered or granular baits to shelves or floors in areas where they can be hazardous to children or pets or cause contamination of food and other items.

Insects may infest poisoned bait if it is left in a bait station for a long time, so replace bait frequently. Remove uneaten bait and thoroughly clean bait stations. Dispose of old or unused bait in an *approved hazardous waste disposal area*; contact the local agricultural commissioner's office for information on toxic material disposal.

When a toxic material is applied to grains and other materials to make poisoned bait, it must be colored with a dye. Coloring serves several useful purposes: (1) it helps avoid mistaken identification so that grains are not used for human or livestock feed; (2) it is a convenient way of identifying the toxicant as specific colors are generally used for certain types of poisons; (3) it makes bait unrecognizable or unattractive to some nontarget organisms; and (4) it provides a convenient way of ensuring that the bait is uniformly treated with the toxicant.

Rodent Bait. For controlling rodents, place bait near nests and along travelways. Rats usually do not go out of their way to find it, and mice confine their activities to small areas most of the time. To improve the chances of it being discovered, place bait in several areas rather than in just a single location. Each of the bait locations for mice, for example, should be no more than 10 feet from another source of bait. Whenever possible, put bait under cover of some object so the rodents feel secure while feeding. Secure bait for roof rats in rafters, trees, or other elevated areas. For Norway rats, place bait along the bases of walls and near ground burrows; it is also possible to place bait in burrows and put a rock or other heavy object over the burrow opening so children or nontarget animals cannot reach it.

Insect Bait. Put insect bait in areas of greatest activity or in areas that cannot be sprayed or dusted. For ants, locate the bait along trails, near nest entrances, around the foundation of the building, and under sinks and other out-of-the-way locations inside the building. Apply cockroach bait under appliances, under sinks, behind furniture, and in hidden areas where these insects have been observed or are suspected to occur; place bait at wall intersections as cockroaches tend to travel along edges. For cockroach species that occur outdoors, place baits around or in woodpiles and in water meter boxes and other protected locations where these insects are usually found.

Bait Stations. Be sure bait stations are suitably designed for the kind of bait being used and the pest being baited. For rodents, use bait stations that comply with the rodenticide label. These should be large enough to accommodate several rats or mice at a time. Provide at least two 1-inch openings into the feeding station for mice and two 2 ½-inch openings for rats. Multiple small openings are important for an insect bait station.

Use only tamperproof bait stations to prevent children, pets, or nontarget animals from gaining access to the bait. Tamperproof refers to a design that blocks access to the bait either through the opening used for filling the station

Tamperproof Bait Boxes

A tamperproof bait box must meet the following criteria to be acceptable for use:

1. Resistant to weather. Placement of the bait station influences weather resistance. If the bait box is placed outside, it needs to be more resistant than if placed indoors or under a shelter.

2. Strong enough to prohibit entry by large, nontarget species. Placement may be a factor if the bait box is inaccessible to nontarget species because it is located inside a building or shelter.

3. Equipped with a locking lid.

4. Equipped with entrances that readily allow target animals access to baits while, at the same time, denying access to larger nontarget species. Access to larger species may be restricted by using baffles, mazes, or small entrances.

5. Capable of being anchored securely so that the bait box cannot be moved or its contents displaced.

6. Equipped with an internal structure for confining the bait. In most boxes, this consists of an arrangement of baffles.

7. Made in such a way so as not to be an "attractive nuisance."

8. Capable of bearing precautionary statements in a prominent location. The bait box must meet service container labeling requirements.

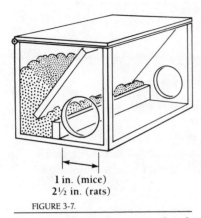

1 in. (mice)
2½ in. (rats)

FIGURE 3-7.

Bait stations must be equipped with an internal baffle, as illustrated, to prevent rodents from scattering the bait.

or through the openings that rodents use. Stations must be secured to a surface to prevent them from being tipped or the toxic bait shaken out. Be sure the word "Poison" is clearly printed on each bait station. Bait stations are considered to be service containers, so they must be labeled with the following information: (1) the name and address of person or firm responsible for the bait station; (2) the identity of the poison being used; and (3) the signal word from the pesticide label. If the bait is a grain or granule, use a bait station equipped with an internal baffle to keep rodents from scattering it (Figure 3-7).

Resistance and Bait Shyness. If baiting has been used before, but control was not successful, the target pest may be developing pesticide resistance or bait shyness.

Pesticide resistance is an acquired condition that gives the target pest population a tolerance or immunity to the toxic substance (Figure 3-8). When resistance is suspected (pests are eating the bait but the population is not declining), switch to another control method such as trapping. Otherwise, use a toxicant that has a different mode of action and augment the baiting program with other methods of control. Discontinue the use of bait for several months to reduce chances of further resistance developing.

Bait shyness develops if an individual animal dislikes the bait or has had a bad experience with it. Bait may be unattractive because it is old, moldy, or contaminated. If pest rodents are not feeding on the bait, check to be sure it is fresh and uncontaminated; use another type of attractant and select a toxicant with a different mode of action to see if this improves acceptance. Be sure bait is located in areas where target pests have access to it. Sometimes prebaiting—setting out the same type of bait minus the toxicant—is helpful in overcoming bait shyness. Once the nontoxic bait is being taken regularly, switch to the toxic bait.

FIGURE 3-8.

This drawing illustrates how pesticide resistance can build up in a pest population. Resistance to pesticides involves a change in the genetic characteristics of pest populations which are inherited from one generation to the next. Increased or frequent use of a pesticide often hastens resistance.

resistant individual

susceptible individual

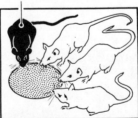

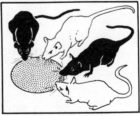

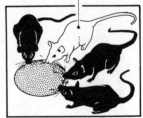

1. Some individuals in a pest population have genetic traits that allow them to survive an application of poison.

2. A proportion of the survivors' offspring inherit the resistance traits. At the next application these resistant individuals will survive.

3. If poisons are applied frequently, the pest population will soon consist mostly of resistant individuals.

HOW PESTICIDES CAN INJURE PEOPLE

Poisonous chemicals such as pesticides injure or kill people by interfering with the normal functioning of internal organs and systems. The nature and extent of injury depends on the toxicity of the chemical as well as the dose (amount of material) that enters the body's tissues. A person's health and size may also influence the severity of injury.

The ingredients of some pesticides are very potent and are capable of causing poisoning at doses as small as a few drops or a few ounces. Other

Avoiding Pesticide Exposure

There are many things you can do to avoid pesticide exposure when working around pesticides or applying them. Follow these guidelines:

- Wear clean protective clothing whenever you work around pesticides or work on application equipment. Wash thoroughly before eating or drinking, smoking, or using the bathroom. Change clothes and bathe after handling or applying pesticides. Check the pesticide label for protective clothing requirements. See Table 3-5 for guidelines in interpreting label requirements.

- Practice safe mixing and application methods. Never walk or drive through freshly treated areas. Clean up leaks and spills as soon as they occur. If any pesticide spills on you, remove contaminated clothing immediately and wash thoroughly with plenty of soap and water.

- Whenever possible, select the least hazardous pesticides. Reduce pesticide use by combining chemical control methods with nonchemical control methods. Apply pesticides only as spot treatments and time the applications to coincide with the most susceptible stage of the target pest.

less-potent pesticides might require that as much as several pounds be consumed before signs of illness appear. Regardless of the specific potential hazard, anyone working with pesticides should avoid exposure by using suitable protective clothing and application techniques. Anyone living or working in pesticide-treated areas must be protected from exposure levels that will cause injury.

Poisoning Symptoms

Symptoms are abnormal conditions, feelings, or signs that indicate the presence of an injury, disease, or disorder. When a person is exposed to a large enough dose of pesticide to cause injury or poisoning, some type of symptoms will usually appear (Table 3-4). These symptoms may show up immediately or after several days; sometimes they may not appear until after several months or years. It may be difficult to associate the illness or injury with its cause if there has been a lapse of time between exposure and observable effect.

The effect of an exposure can be localized, such as eye or skin irritation, or generalized, when the pesticide is absorbed into the blood and distributed to other parts of the body. A pesticide can affect several different internal systems at the same time. If the person experiences an injury but recovers quickly, or gets worse and dies within a short time, it is known as an acute illness or injury. If the effects last for a long time, and perhaps are irreversible, it is known as a chronic illness. Examples of chronic illnesses usually associated with high or prolonged levels of exposure to certain pesticides include, among others, infertility, birth defects, and cancer. Pesticides that are found to cause such disorders or are suspected of causing these problems may lose their federal registration and can then no longer be used in the United States.

TABLE 3-4

Common Pesticide Poisoning Symptoms.

POSSIBLE SYMPTOMS RELATED TO SKIN CONTACT
WITH PESTICIDE DUST, LIQUID, OR VAPORS

   Staining of the skin

   Reddening of skin in area of contact

   Mild burning or itching sensation

   Painful burning sensation

   Blistering of the skin

   Cracking and damage to nails

   Involvement of internal systems resulting in blurred vision, dizziness, vomiting,
   or diarrhea

   Possible muscle weakness, poor coordination, muscle cramps

   Potential chronic problems (see below)

POSSIBLE SYMPTOMS RELATED TO EYE CONTACT
WITH PESTICIDE DUST, LIQUID, OR VAPORS

   Discomfort, including watering and slight burning

   Severe, painful burning (permanent eye damage may occur)

   Involvement of internal systems resulting in blurred vision, dizziness, vomiting,
   or diarrhea

   Possible muscle weakness, poor coordination, muscle cramps

   Potential chronic problems (see below)

POSSIBLE SYMPTOMS RELATED TO INHALING OR SWALLOWING
PESTICIDE DUST, LIQUID, OR VAPORS

   Sneezing

   Irritation of nose and throat

   Nasal stuffiness

   Swelling of mouth or throat

   Coughing

   Breathing difficulties

   Shortness of breath

   Chest pains

   Involvement of internal systems resulting in blurred vision, dizziness, vomiting,
   or diarrhea

   Possible muscle weakness, poor coordination, muscle cramps

   Potential chronic problems (see below)

CHRONIC PROBLEMS

Exposure to some types of pesticides may result in chronic problems such as cancer,
infertility, birth defects, or genetic disorders to the exposed person or their offspring. Symptoms of these disorders may not appear until years after exposure. Repeated exposure to
low doses of certain pesticides over long periods of time may increase the potential for
chronic health problems; single incidents of high-level exposure to certain pesticides may
also increase the possibility of chronic health problems.

If the evidence of chronic or acute health hazards warrants, the California
Department of Food and Agriculture may prohibit the use of certain pesticides
within the state even though they still may have a valid federal registration.

   Some pesticide poisoning symptoms are similar to symptoms produced by
many other chemicals. The type of symptoms may vary between chemical
classes of pesticides and may also be different among pesticides within the
same chemical class. The presence and severity of symptoms usually are
proportional to the amount of pesticide (the dosage) entering the tissues of

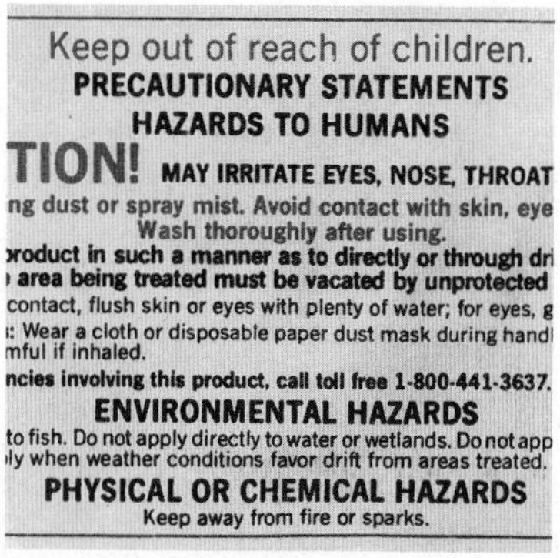

Keep out of reach of children.
PRECAUTIONARY STATEMENTS
HAZARDS TO HUMANS
TION! MAY IRRITATE EYES, NOSE, THROAT
ng dust or spray mist. Avoid contact with skin, eye
Wash thoroughly after using.
product in such a manner as to directly or through dri
area being treated must be vacated by unprotected
contact, flush skin or eyes with plenty of water; for eyes, g
: Wear a cloth or disposable paper dust mask during handl
mful if inhaled.
ncies involving this product, call toll free 1-800-441-3637.
ENVIRONMENTAL HAZARDS
to fish. Do not apply directly to water or wetlands. Do not app
ly when weather conditions favor drift from areas treated.
PHYSICAL OR CHEMICAL HAZARDS
Keep away from fire or sparks.

FIGURE 3-9.

Follow the pesticide label use instructions carefully and observe any precautions listed.

the exposed person. Symptoms may include a skin rash, headache, or irritation of the eyes, nose, or throat; these types of symptoms may go away within a short period of time and sometimes are difficult to distinguish from symptoms of an allergy, cold, or the flu. Other symptoms, which might be caused by higher levels of pesticide exposure, include any of the following: blurred vision, dizziness, heavy sweating, weakness, nausea, stomach pain, vomiting, diarrhea, extreme thirst, and blistered skin. Poisoning can also result in apprehension, restlessness, anxiety, unusual behavior, shaking, convulsions, or unconsciousness. Although these symptoms can indicate pesticide poisoning, they also may be signs of another physical disorder or disease. Whenever the possibility of poisoning exists, consult a physician; be sure to give the physician a copy of the pesticide label or the name of the pesticide, the manufacturer, and the EPA registration number. Diagnosis of a pesticide-caused injury usually requires careful medical examination, laboratory tests, observation, and familiarity with a person's medical history.

Individuals commonly vary in their sensitivity to pesticides. Some people show no reaction to a dose that causes severe illness in others. A person's age and body size may influence their response to a given dose, thus infants and young children are normally affected by smaller doses than adults. Also, adult women may be affected by smaller doses of some pesticides than adult men. The unborn child carried by a pregnant woman may be highly sensitive to exposure to some pesticides.

Pesticides that are applied in strict accordance with their label instructions—with adherence to application rates, reentry intervals, protective equipment requirements, aeration periods, and other listed procedures—generally do not leave unsafe levels of pesticide residues (Figure 3-9). Accidents during application may result in a higher, and sometimes unsafe, exposure. An improper application caused by not following label instructions may also result in injury.

PROTECTING PEOPLE

Always apply pesticides in strict accordance with label instructions. Furthermore, never use a pesticide in a building or other area unless people living or working there can be protected from exposure. This usually requires that they leave the area before an application begins and that they remain away for a period of time after the application has been completed. Provide occupants with information about the pesticide application and be sure they understand what safety precautions are being taken. The type of information they may need includes (1) the name of the material being used; (2) poisoning symptoms and what to do if they experience such problems, where to get help, and how to get more information; (3) what areas of the building are being treated; (4) what to expect, such as an odor or residue; and (5) the possibility of finding dead insects or rodents and what to do if this happens. Explain ways to reduce personal exposure, such as removing or covering food and utensils before pesticide applications are made, protecting linens and bedding and similar items, opening windows and doors to increase ventilation after an application has been made, vacuuming carpets and cleaning floors after an application, and keeping children and pets away from treated areas.

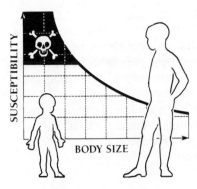

FIGURE 3-10.

Body size may be an important factor in determining an individual's susceptibility to a pesticide. Usually infants and children are more susceptible and may be injured by smaller doses.

Pesticides may be needed to control pests in places where food is stored, prepared, or eaten. If so, special precautions must be taken. For instance, never treat food preparation surfaces with dusts or liquid sprays and do not allow residues to drift onto food or utensils. If fogs are used, all food preparation surfaces must be thoroughly cleaned after application.

Never make an application near air ducts or ventilation systems unless the system can be shut down for a period of time. Do not apply pesticides inside heating or cooling ducts.

Infants, Children, the Elderly, and People with Medical Conditions

Sometimes the use of pesticides in buildings must be restricted or avoided to protect people living there. Rely on nonchemical control methods as much as possible and use a pesticide only where absolutely necessary. When pesticides are needed, choose the safest formulation available such as a bait or a liquid spray having low volatility. Follow label instructions and precautions carefully. Be extremely careful when using pesticides in areas occupied by infants, children, the elderly, or someone who is sick. These areas include hospitals, nursing homes, schools, and certain households.

Infants are more vulnerable to pesticide exposure than larger children or adults (Figure 3-10). This is because of their small size and undeveloped immune system responsible for detoxifying hazardous chemicals. Do not apply a pesticide to anything used for infant care, and avoid spraying or dusting carpets, clothing, blankets, towels, or any fabrics that infants or others may contact. When a pesticide is needed in areas where an infant may spend part of the day, use a formulation that will break down completely before the infant returns.

Children under the age of six are active and curious. It is difficult to keep them away from places where a pesticide has been used for control of household pests (Figure 3-11). Young children do a lot of exploring and put many objects (including their hands) into their mouths. They also crawl on floors and climb on other surfaces. Therefore, never apply a pesticide to play equipment, toys, or any surfaces normally contacted by the youngsters. On carpets, use pesticides that break down rapidly. In all cases, use pesticides having low toxicity and low volatility. If you use bait stations or traps, secure them well out of reach and out of sight.

FIGURE 3-11.

Young children are very active, making it difficult to keep them out of areas where pesticides may have been applied. Use extreme caution if pesticides must be used in areas where young children may live or play.

Elderly people may be susceptible to respiratory illnesses and other disorders that may give them a low tolerance to many airborne dusts and chemicals, including specific pesticides. In some instances, their bodies may not be able to properly degrade or eliminate foreign or toxic materials, such as pesticides. Therefore, use extreme caution when making pesticide applications in rooms where elderly people sleep or spend long periods of time; whenever possible, avoid treating these places. In other areas, use a low-toxicity and low-volatility pesticide. Apply this as a spot treatment only as necessary. Select alternate methods of control whenever possible, and always augment pesticide use with other pest control techniques so that the amount of pesticide used can be reduced.

People who are acutely ill or suffer from conditions such as diabetes or alcoholism or have allergies or respiratory disorders including asthma and emphysema may be more sensitive to pesticides in their environment. Medications used to treat illnesses may influence the effects of pesticide exposure. Provide persons who are ill or using medications with the name of the pesticide you plan to use and ask them to contact their physician for advice.

APPLICATOR SAFETY

Safety risks for applicators working in buildings or enclosed areas are compounded by hazards such as electrical equipment, possibility of explosions, and confined work areas. Learn to recognize hazards in the application site that could cause injury. Avoid pesticide exposure by wearing required or recommended protective equipment. Table 3-5 gives examples of suitable protective equipment based on label recommendations. Maintain, clean, and store protective equipment carefully to keep it in good condition and to ensure that it provides optimum protection (Table 3-6).

Fire, Explosion, and Electrical Hazards

Fires, explosions, and electrical hazards can be found in residential, industrial, and institutional settings and other confined areas. Before using a pesticide, examine the application site for hazards. For example, never apply a pesticide dissolved in oil or petroleum solvent in an enclosed area if there is any source of spark or flame such as functioning electrical motors, wall switches, appliances, or pilot lights; before making an application, shut off electric and gas services to the treatment area. Avoid the use of aerosols in wall voids near hot water pipes; heat from these pipes can ignite solvents and cause a fire. Do not use dust in an enclosed area if there is an ignition source

PROTECTIVE CLOTHING AND EQUIPMENT REQUIREMENTS

 A daily change of clean coveralls or clean outer clothing. Wear waterproof pants and jackets if there is any chance of becoming wet with spray. Disposable suits of Tyvek can be used in some, but not all situations. Uncoated Tyvek can be worn in place of coveralls or long sleeved shirt and pants. It will not take the place of waterproof outer clothing. Tyvek which has been coated with polyethylene can be worn in place of waterproof clothing in some situations, but not with organophosphate liquids. The solvents in these pesticides will break down the polyethylene coating. Saranex coated Tyvex suits can be used effectively with organophosphates. Neither uncoated or Saranex coated Tyvek adequately protect against chlorinated hydrocarbons such as methyoxychlor.

 Waterproof apron made from rubber or synthetic material. Use for mixing liquids.

 Waterproof boots or foot coverings made from rubber or synthetic material.

 Faceshield. goggles, or full face respirator. Goggles with side shields or a full face respirator is required if handling or applying dusts, wettable powders, or granules or if being exposed to spray mist. Safety glasses with brow and temple protection may be worn if the label does not specify goggles or face shield.

 Waterproof, unlined gloves made from rubber or synthetic material.

 Waterproof, wide-brimmed hat with nonabsorbent headband, or a hood if wearing a waterproof plastic rainsuit with hood attached.

 Cartridge type respirator approved for pesticide vapors unless label specifies another type of respirator such as a dust mask, canister type gas mask, or self-contained breathing apparatus.

TABLE 3-5

Protective Equipment and Clothing Guide.

SUMMARIZED LABEL STATEMENT *Toxicity Category*	MIXER-LOADER		APPLICATOR	
	I-II	III	I-II	III**
Precautions should be taken to prevent exposure.	A B C F G H R *	A B C F G H	B C F G H R *	C F G H R *
Protective clothing or protective equipment is to be worn.	A B C F G H R *	B C F G H R *	B C F G H R	C F G H R *
Clean clothing is to be worn.	C	C	C	C
Contact with clothing should be avoided.	A B C	B C	B C	C
Contact with shoes should be avoided.	B	B	B	B
Rubber boots or rubber foot coverings are to be worn.	B	B	B	B
Contact with skin should be avoided.	A B C F G H	B C F G H	B C F G H	C F G H *
A cap or hat is to be worn.	H	H	H	H
An apron is to be worn.	A	A		
Rubber gloves are to be worn.	G	G	G	G
Contact with eyes should be avoided	F	F	F	F
Goggles or faceshield is to be worn.	F	F	F	F
Avoid inhalation.	R	R	R *	R *
A respirator is to be worn.	R	R	R *	R *

*Use this equipment when there is a likelihood of exposure to spray mist, dust, or vapors.

**If the Category III pesticide application is being made in an enclosed area such as a greenhouse, or if the application consists of a concentrate spray of 100 gallons-per-acre or less in a grove, orchard, or vineyard, then use the protective equipment guidelines for Category I-II pesticides.

TABLE 3-6

Selecting and Maintaining Personal Protective Equipment.

ITEM	USES/PROBLEMS	MAINTENANCE
PROTECTIVE EYEWEAR		
Goggles	Suitable for most mixing and application jobs. Lenses scratch easily. May fog up. Choose goggles with nonabsorptive head band.	Clean daily with soap and water. Replace scratched lenses and worn straps.
Safety glasses	Must have brow and side shields. More comfortable than goggles. Do not provide as much protection as goggles. Available with tinted lenses.	Clean daily with soap and water. Replace when lenses become scratched.
Faceshields	Suitable for mixing but not for most application situations. Scratches easily. Must have nonabsorbent headband.	Clean daily with soap and water. Replace when plastic faceshield becomes scratched.
PROTECTIVE HEADWEAR		
Plastic hard hat	Must have nonabsorbent head band.	Clean daily with soap and water.
Hood on waterproof jacket	Should not be removable. Must be unlined.	Clean daily with soap and water.
PROTECTIVE CLOTHING		
Woven long-sleeved shirt and long pants	Minimal protection but should be worn when more protection not required. Avoid wetting with liquid sprays.	Launder daily with hot water and liquid detergent.
Woven coveralls	Minimal protection. Can be removed easily if contaminated. Protects clothing underneath. Avoid wetting with liquid sprays.	Launder daily with hot water and liquid detergent.
Disposable coveralls	Several types offering different types of protection. Unlaminated materials offer similar protection as woven materials. Laminated materials offer protection similar to waterproof materials.	Generally not reused. Throw away after each day or launder with hot water and liquid detergent.
Waterproof rainsuit	Maximum protection. Must have attached hood. Must be unlined or have nonabsorbent lining.	Launder daily with hot water and liquid detergent. Check daily for cracks and tears.
Waterproof gloves and boots	See chart below for suitable materials. Gloves should be unlined.	Wash gloves daily with mild soap and water. Check daily for holes and cracks.

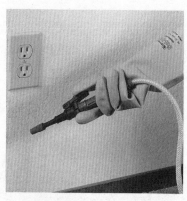

FIGURE 3-12.

To avoid chances of electrical shock, do not apply water-based sprays around electrical outlets or appliances.

as any airborne dust at the right concentration can explode. Boric acid dust is capable of extinguishing a pilot light, which could create an explosion hazard due to escaping gas (most new gas appliances are equipped with safety shut-off devices or igniters in place of pilot lights).

Do not use a water-based spray around electric appliances, outlets, or switches unless the power has been shut off. Water conducts electricity, so you risk electrocution if the spray touches a live power source (Figure 3-12).

Power tools and other electrical equipment that you may use during a pest control operation can also create hazards. Wiring in older buildings may not accommodate heavy-duty electrical equipment. Before connecting equipment, use a circuit tester to make sure the outlet is correctly grounded. Check the wiring size and the fuse or breaker box to be sure that the system can handle the electrical demand of the equipment being used. If the circuit is not protected with the correct size fuse or circuit breaker, or if wiring is too small, an overload could heat the wiring and start a fire. Inadequate grounding can cause a fatal electric shock; prevent this hazard by using a ground fault interrupter (GFI).

FIGURE 3-13.

Reduce exposure hazards when working in confined areas by wearing personal safety equipment.

Working in Confined Areas

Confined areas present special hazards to persons making a pesticide application. Confined areas may be attics, crawl spaces beneath buildings, storage areas, closets, small rooms, and other places that have poor ventilation. Hazards include inhaling the pesticide being applied and coming in contact with treated surfaces. Cramped areas may be uncomfortably hot due to poor air circulation. High temperatures may increase your exposure potential, because sweating and high temperatures accelerate the rate of skin absorption of some pesticides.

Reduce exposure hazards when working in confined areas by wearing personal safety equipment (Figure 3-13). Whenever possible, increase ventilation in the treatment area by opening windows or using a fan to bring in fresh air. Always begin the application from a point furthest from the exit; never walk or crawl through freshly applied pesticide.

To avoid breathing fumes, wear an approved pesticide respirator. Be sure it seals well around your face and is in good working condition. A cartridge or canister type respirator must be worn whenever a Category I or Category II pesticide is being used in confined areas. Applicators having beards or long sideburns must use a powered cartridge respirator, because facial hair prevents adequate sealing of conventional respirator face masks. When atmosphere monitoring equipment indicates that an oxygen deficiency condition exists, or when applying a fumigant, a supplied-air respirator is required.

Prevent skin or eye contact with spray residue or vapor. Always wear a long-sleeved shirt and full-length pants, coveralls, or lightweight spray suit when making an application. Protect your hands with waterproof gloves and use a faceshield or goggles to prevent spray or dust from getting into your eyes. Check the pesticide label for the minimum protective clothing requirements.

Protecting Pets and Domestic Animals

Pets housed in or near residences or other buildings include several types of mammals, birds, reptiles, amphibians, and fish. Associated with pets and

Technique for Properly Fitting Cartridge Respirators

ISOAMYL ACETATE
(banana oil) FIT TEST*

The chemical isoamyl acetate, commonly referred to as "banana oil," is available from major chemical suppliers and is widely used to check respirator fit. Its odor is easy to detect and the chemical can be used with any pesticide respirator equipped with organic vapor cartridge or canister.

When conducting a fit test, it is important to know that some brands of respirators are available in small, medium, and large sizes. If possible, have several different sizes available during the test to ensure proper fit. Try respirators from different manufacturers since one brand may fit better than others.

If a respirator does not fit properly, the applicator will not be adequately protected. Therefore, be sure to follow the test procedures outlined below:

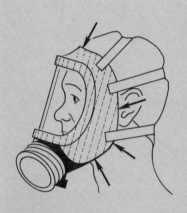

1. Be sure there is no banana oil odor in the test area that may influence the wearer's ability to detect its presence. Once a respirator is selected, have the wearer adjust it until thre is a good face-to-mask seal.

2. Saturate a piece of cotton or cloth with banana oil. The person performing the test should wear rubber gloves and avoid skin contact with the wearer.

3. Pass the saturated material close to the respirator in a clockwise and counterclockwise motion. Have the wearer stand still and breathe normally and then deeply. If the wearer smells banana oil, readjust the respirator or select a different size or style before starting again.

4. If the odor cannot be detected while the wearer is standing still, have them perform side-to-side and up-and-down head movements. Also have the wearer talk loudly enough to be heard by someone standing nearby. Then have the person make other movements, such as bending over, that may occur during spray application.

5. If the banana oil odor cannot be detected during the above movements, it indicates a satisfactory fit. Seal the respirator in a plastic bag marked with the wearer's name. Keep a record of when the fit test was conducted, along with the size and brand of respirator selected for each user.

*Adapted from *A Guide to the Proper Selection and Use of Respirators,* Zoecon Corporation.

domestic animals are their food and water supplies, bedding, pens, equipment, and toys.

Most animals are susceptible to injury by pesticides, even some types that are applied at low doses. Fish and birds are among the most susceptible. Cats are very sensitive because they are metabolically unable to detoxify many types of pesticides. Young animals and older or sick animals may be affected by lower pesticide doses than adult or healthy animals. Cats and dogs lie and sleep on the ground and other surfaces that may have been treated. They groom and clean themselves by licking, which increases their potential for exposure even when small amounts of pesticide have been used.

To provide protection for pets and domestic animals, remove them from the area before making a pesticide application. Keep animals away until the spray dries and the area is well ventilated. Do not apply pesticides on or near animal food or water. If the animals are returned to the treated area, flea collars should be removed; any ectoparasite systemic medications should be discontinued.

Pets or domestic animals can be the source of some pest problems. For instance, fleas are usually brought into a building by dogs or cats; dogs may also carry in ticks. Animal manures provide food and breeding sites for several fly species. Pet or livestock food or food left in an animal dish or feeder can attract mice and rats as well as cockroaches, flies, and ants. An animal's water dish may provide the water needed by some pests. Therefore, when performing pest management in an area where pets or livestock are kept, look for these types of conditions. Evaluate and, if necessary, suggest modifications of the feeding routine, housing arrangement, and sanitation practices to reduce pest problems.

Pesticide Drift

If pesticides are not carefully applied, they may drift beyond the treatment site and become deposited as unacceptable residues on surfaces not intended to be treated. These residues can possibly endanger nontarget organisms. Residues from improper application or improper rinsing of equipment may also result in contamination of surface or groundwater.

Preventing Drift or Unwanted Exposure

Do not use dusts in outdoor locations. To prevent drift when applying liquid sprays, use low pressures and large nozzle orifices. This prevents formation of small droplets subject to drift. Never make an outdoor application of a liquid spray when the wind is blowing faster than 5 miles per hour. If there is a slight wind, select a formulation or adjuvant that reduces drift. Be especially careful if you are spraying near fruit trees or vegetable gardens, flowers, laundry being air dried, cars, windows, dark surfaces that may spot, pet or livestock food and water containers, fish ponds, bird baths, swimming pools, saunas, spas, or outdoor furniture. Avoid outdoor applications that may drift to children's play areas, sandboxes, swing sets, or lawns and shrubbery that children contact.

Do not apply a pesticide in outdoor locations where residues can be carried into a well, stream, pond, or other water source. Never drain or wash application equipment where runoff into sewers, sinks, sumps, or drainage tiles can occur.

When making a liquid or dust application inside a structure, keep the spray or dust away from air ducts, fans, or blowers to prevent the material from being blown around.

CHARACTERISTICS OF TREATED SURFACES

Treatment sites may have surfaces whose characteristics must be evaluated before applying a pesticide. Depending on the type of surface, a pesticide can be absorbed and rendered ineffective, or the surface may be stained or etched. Concrete, for example, is porous and tends to absorb liquid sprays, reducing the amount of residue on the surface that is available to control target pests.

Floor coverings such as linoleum, tile, and carpeting can be stained or etched by some pesticides or solvents. Certain wallpapers and carpets contain dyes that may run, dissolve, or change colors if exposed to components of some pesticides. Paint and other finishes used on walls or woodwork may also react with these chemicals to produce spotting or discoloration. Fabrics of all types, and the dyes used for patterns and color, may also react, affecting wear or causing a stain or change in color. A soiled fabric may react differently than a clean one. Fabrics also can absorb a liquid pesticide, reducing pest control effectiveness.

Dust formulations leave an unsightly residue if applied to surfaces of furniture, woodwork, fabrics, and other items in the treatment area.

Preventing Problems

Stains or color changes may be caused by an excessive dose or by certain application techniques. The formulation type may affect staining or spotting. A soiled or greasy surface may increase staining, spotting, or absorption. Paint that has been recently applied and not fully cured has more of a tendency to spot.

Whenever possible, first apply a pesticide to an inconspicuous area, such as a closet, and allow the pesticide to dry for several hours to observe the reaction. Be careful when treating upholstery, furniture, drapes, or lower wall surfaces with a pesticide (lower wall surfaces are more likely to be soiled, which may enhance staining or bind the pesticide to make it less effective). Read and follow label directions and precautions carefully to avoid staining, spotting, visible residues, and pesticide deactivation. Thoroughly clean the application equipment before adding a pesticide to prevent a possible reaction between the pesticide and contaminants in the equipment. These contaminants may cause stains or other adverse effects.

When two or more pesticides are mixed, additional problems associated with pesticide compatibility may appear. Check the compatibility of pesticide mixtures before application.

Odor Problems

Many pesticides have odors that can be detected during and after application. Odors are usually strongest when pesticides are first applied. In confined areas, odors may become overpowering and objectionable; they can cause nausea or headache, initiate asthma or other breathing difficulties, or trigger other medical or anxiety-related symptoms.

An odor may be a chemical characteristic of the pesticide or its solvent, or it may be a substance added to the pesticide as a warning agent to reduce chances of injury. Reduce problems associated with odors by (1) using only the application rate stated on the pesticide label, (2) applying the pesticide in localized areas or as a spot treatment whenever possible, (3) using a low-odor formulation if available and if appropriate, (4) increasing ventilation to the application area by opening windows and doors or using fans, and (5) applying the pesticide during periods when the building is not occupied.

An odor may also be caused from a reaction between the pesticide and surfaces that have been treated. Before applying any pesticide in a confined area, read the pesticide label to determine if any of the chemicals in the formulation will react with treated surfaces to produce an odor.

TRANSPORTING PESTICIDES

Pesticides must be transported with special care to prevent spills or accidents that might possibly injure people and animals or damage the environment. A pesticide spill on a roadway can result in serious problems.

Pest control service vehicles such as pickup trucks or vans are generally used to carry pesticides and application equipment to work sites (Figure 3-14). Some pesticides may be in original containers or service containers; others may be in a spray tank or application device. No matter what form they are in or how they are contained, pesticides transported on public roads are classified by regulatory agencies as hazardous materials. Unused spray material may be classified as a hazardous waste. Classification as a hazardous waste greatly complicates the manner in which pesticide materials can be transported, stored, and disposed of.

Governmental agencies regulate hazardous material and hazardous waste transportation on public roads. Under certain conditions, a permit may be required to transport hazardous materials or wastes. Transportation regulations also require that certain vehicles be equipped with placards indicating the class of hazardous material being carried. Vehicles may be subject to inspection by the California Highway Patrol. See Table 3-7 for important factors that should be considered when transporting pesticides in a vehicle. Consult with the California Highway Patrol and the California Department of Transportation for information on regulations and permits.

During transport, keep undiluted pesticides in their original containers or in approved, labeled service containers. If a container has previously been opened, be sure it is tightly resealed before transporting. Carry diluted pesticides in approved containers and label them according to state and federal regulations. Application equipment or service containers containing pesticides must be labeled with the name of the pesticide, the toxicity signal word from the original container, and the name and address of the person or company responsible for the container (Figure 3-15). It should also bear the statement "KEEP OUT OF THE REACH OF CHILDREN."

During transport, secure all pesticide containers and application equipment to avoid spills or container damage. Use sand bags, blocks, ropes, or straps

FIGURE 3-14.

Pest control service vehicles such as pickup trucks or vans are generally used to transport pesticides and application equipment to work sites.

TABLE 3-7

Factors to Consider When Transporting Pesticides.

CONTAINERS

Use original container. Be sure container is sealed. Use approved service container, tightly sealed. Use application tank or equipment with proper seal.

LABEL

All containers and application equipment must be labeled to show contents of container, signal word, responsible party, and the statement "Keep Out of Reach of Children."

VEHICLE

Transport pesticides in a truck where cargo is separate from passenger area. Do not carry people or animals in cargo area. Do not carry food or animal feed in cargo area. Secure containers and equipment containing pesticides. Do not stack pesticide containers higher than sides of vehicle's cargo area.

PLACARDS

Placards may be required on vehicle. Check with California Highway Patrol or Department of Transportation. If required, placards must be placed on all four sides of vehicle and be clearly visible.

SECURING VEHICLE

If vehicle is ever left unattended, pesticides must be secured in a lockable container. Covers on tanks containing pesticides must be locked or equipment must be in a locked part of the vehicle.

ACCIDENTS

Accidents involving pesticide spills on public roads must be reported to local police and fire authorities or the California Highway Patrol immediately. Call "911." Never leave the accident site until another responsible party arrives and supervises the cleanup. Keep people away from the spill.

RECORDS

Keep records of all pesticides carried in the vehicle. Have copies of the Material Safety Data Sheets for each pesticide in the vehicle. This information is useful to emergency workers in the event of an accident.

PROTECTIVE COTHING AND EQUIPMENT

Do not wear or store contaminated clothing or equipment in the passenger compartment. Store clean clothing and equipment in a separate compartment from contaminated clothing and equipment.

FIGURE 3-15.

Be sure application devices are labeled with the name of the pesticide, its EPA registration number, the name of the active ingredient and its weight or percent in the formulation, the toxicity signal word, and the name, address, and telephone number of the person or company responsible for the container.

to prevent movement. The vehicle should be equipped with an emergency spill control kit, including a supply of absorbent material, a special container for holding waste, and a quantity of clean water. If a spill occurs, no matter how small, clean it up immediately. Table 3-8 is a summary of pesticide cleanup.

Lock the area within the vehicle where pesticides are carried to keep children or unauthorized adults out when the vehicle is unattended (Figure 3-16). Also, lock tanks containing diluted pesticides, or store tanks and other equipment containing pesticides in a locked area on the vehicle that is separate from food, feed, or passengers.

TABLE 3-8

Steps to Follow in Cleaning Up a Pesticide Spill.

1. Wear protective equipment, including rubber boots, gloves, waterproof protective clothing, goggles, and respiratory protection.

2. Clear the area and prevent unprotected people from coming near the spill.

3. Administer first aid and obtain medical care for anyone who received a pesticide exposure.

4. Prevent fires by extinguishing sources of ignition and providing adequate ventilation.

5. Contain the leak. Use sand or other absorbent to keep the pesticide confined. Patch the leaking container or transfer its contents to a sound container.

6. Clean up pesticide and absorbent and any contaminated objects. Place these materials into a sealable holding container.

7. Decontaminate the area contacted by the pesticide, using a suitable decontamination solution.* Transfer residues to the holding container.

8. Label containers holding spilled pesticide and contaminated soil and other objects. Include pesticide name, signal word, and name of responsible party.

9. Transport holding containers to approved Class I disposal site.

*See Volume 1 (*The Safe and Effective Use of Pesticides*) for information on decontamination solutions for pesticide cleanup.

FIGURE 3-16.

Keep pesticides and application equipment locked in a designated part of the vehicle to prevent children or unauthorized adults from gaining access to them.

4 Weed Control

The control of weeds around structures may be an important part of a structural pest control program. For this reason, persons licensed in the Residential, Industrial, and Institutional Pest Control category must be familiar with the principles of nonchemical and chemical weed control. However, if you perform extensive amounts of weed control beyond the perimeters of buildings or are involved in landscape maintenance around buildings, you should consider expanding your applicator certificate or license to include the category *Landscape Pest Management*.

Weeds of major importance possess special characteristics that distinguish them from the occasional out-of-place plant. They adapt well to local climates, soils, and other external conditions and can compete successfully with

TABLE 4-1

Persistence of Weed Seeds in the Soil. A special characteristic of many weeds is the persistence of their seeds. The following illustrates the percentage of seed from seven species of weeds which germinated after 38 years.

WEED SPECIES	SEED GERMINATED AFTER 38 YEARS
Jimsonweed	91%
Mullein	48
Velvet leaf	38
Evening primrose	17
Lambsquarters	7
Green foxtail	1
Curly dock	1

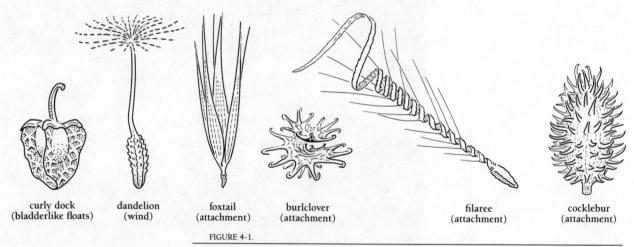

| curly dock (bladderlike floats) | dandelion (wind) | foxtail (attachment) | burlclover (attachment) | filaree (attachment) | cocklebur (attachment) |

FIGURE 4-1.

Weed seeds such as those shown here have many adaptations enabling them to disperse. These special characteristics are among some of the ways weeds compete with other plants—and also why weeds can be found growing in unusual places such as rain gutters.

cultivated plants for available resources. Most weeds produce large quantities of seeds, even under adverse conditions. Seeds of some weeds can lie dormant in the soil for extended periods, sometimes 20 or more years, before germinating (Table 4-1). Some weed seeds or fruits are specially adapted to promote dispersal, which is why weeds can be found growing in unusual places around structures such as in cracks of pavement and in rain gutters (Figure 4-1). Because many weeds are capable of reproducing through vegetative structures, cultural activities such as hoeing or mowing may result in the production of new plants and worsen the weed problem. Consequently, weeds can be persistent and difficult to eliminate.

Other plants that do not have all these characteristics may be considered to be undesirable vegetation if they interfere with some aspect of the utility of a building or its surrounding area. In this case, the undesirable vegetation is considered to be weeds. Grasses, flowering plants, shrubs, and even trees may need to be controlled because of problems they are causing. For instance, tree roots buckling concrete create unsafe conditions (Figure 4-2).

FIGURE 4-2.

Sometimes trees become undesirable plants if they are causing hazards or growing in the wrong location. Roots from this tree are causing the concrete curbing to buckle.

How Weeds Are Pests

Weeds can cause serious problems around structures because (1) they provide a harborage for pests such as insects, rats, and mice; (2) they interfere with normal building use and maintenance and may detract from the building's appearance; (3) they grow in cracks of pavement and contribute to the breakdown of paving materials; (4) they interfere with other types of pest control operations around structures; (5) they restrict air circulation, light, and other factors needed to keep pest problems from occurring; (6) dried weeds are fire hazards and violate city codes; (7) pollens from certain weeds and ornamental plants bother people with allergies; and (8) some weeds and ornamental plants are poisonous to people, pets, and livestock.

CONTROLLING WEEDS

Successful weed control depends on an accurate identification of the weed species and an understanding of their life cycles. By knowing the common names or scientific names of weeds, you will be able to look up important information about them. For instance, many different herbicides are available for various types of weed control. However, to select the correct herbicide, you need to know if the primary weed problems are grasses, broadleaves, or sedges (Figure 4-3). You must also determine if the weeds are annuals (1-year life cycle), biennials (2-year life cycle), or perennials (live more than 2 years). Finally, to achieve effective control, you sometimes must know when the weed seeds germinate and how local climates and environmental conditions affect germination and plant growth. When preemergent herbicides are used, for instance, the application must be made before weed seeds germinate or it will be ineffective.

The following examples illustrate the importance of accurate weed identification and of understanding the characteristics of weeds.

Example 1. ***Proper Identification:*** Oxalis and clover are very similar in appearance, yet chemical controls are entirely different (Figure 4-4). Selecting the correct herbicide requires that you be able to distinguish between these two different weeds.

Example 2. ***Proper Identification:*** Crabgrass (an annual) and dallisgrass (a perennial) look similar when young and one is commonly mistaken for the other (Figure 4-5). Crabgrass is easily controlled with mechanical methods (such as hoeing) or by treating it with pre- or postemergent herbicides. However, established dallisgrass clumps are difficult to control with a hoe and are not affected by preemergent herbicides. Without proper identification, the wrong control methods might be used resulting in poor or ineffective control and a waste of money and time.

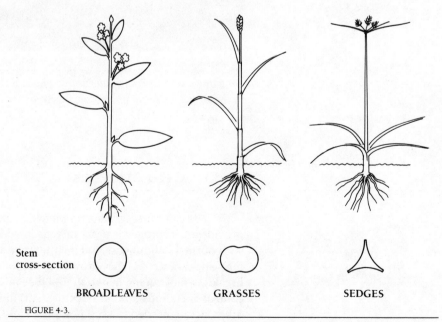

Stem
cross-section

BROADLEAVES   GRASSES   SEDGES

FIGURE 4-3.

Three major groups of plants include the broadleaves (left), grasses (center), and sedges (right). It is important to be able to distinguish these three groups as a first step in identifying weeds.

FIGURE 4-4.

Oxalis (left) and clover (right) are very similar in appearance, yet chemical controls are entirely different. It is important that you learn how to recognize the differences between these plants to be able to select the correct chemical control method.

FIGURE 4-5.

Crabgrass (left) and dallisgrass (right) look quite similar when the plants are young. Control methods are very different, however, so it is important to be able to distinguish these weeds from each other.

Example 3: Proper Identification to Understand the Weed's Life Cycle:
Most annual weeds can be controlled by hoeing the plants slightly below ground level. Perennials are not controlled by this method, however, except through repeated hoeing (which eventually causes carbohydrate starvation). Purslane is an annual and therefore is easily controlled by hoeing. Field bindweed (Figure 4-6), however, is a perennial and is not controlled by hoeing or other mechanical methods. In this case, knowledge of the weed's life cycle is needed to select the correct control measures. The weed must be properly identified to get this information.

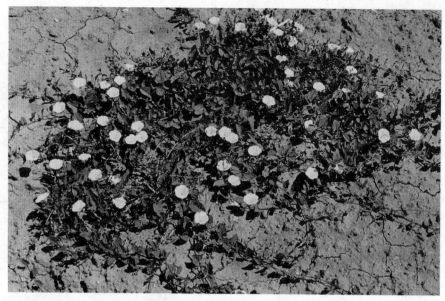

FIGURE 4-6.

Understand the life cycles of weeds before selecting methods to control them. For instance, the field bindweed shown here is a perennial and is very difficult to control with cultural methods such as hoeing.

Example 4: Proper Identification to Understand the Effects of Local Climatic Conditions on Germination: Some weed species growing in different regions of California germinate at different times of the year due to local climatic conditions and weather patterns. For instance, in California's central valley, crabgrass may come up between February 1 and 15 whereas in coastal and cooler climates, it comes up a month later. Knowing the germination time is critical for timing control with preemergent herbicides because these chemicals have to be applied before weed seeds begin to sprout.

Identifying Weeds

A simple way to begin to identify common weeds found around structures is to compare specimens with color photographs and drawings. If you are unable to determine the species from these sources, it may be necessary to use identification keys or to compare the weeds with preserved specimens. Sometimes weed plants need to be sent to an expert for identification (Table 4-2). Advisors in the University of California Cooperative Extension offices, located in counties throughout the state, can assist in identifying weed species.

Experts identify plants, including weeds, by recognizing differences and similarities between flowers, leaves, stems, and roots; fruits, seeds, and special structures are also useful characteristics. Sometimes a plant's growth habits assist in identification.

Identification of a weed by its accepted common name, as opposed to its scientific name, is usually sufficient. If needed, scientific names can be determined by using a weed identification manual that has a cross-reference to common names. For chemical control methods, most herbicide labels list weeds by both common and scientific names.

TABLE 4-2

Guidelines for Sampling and Sending Weeds for Weed Identification.

SAMPLING

1. Choose several plants that represent the species.
2. Include stems, leaves, flowers (if present), and roots.
3. Dig up weeds to prevent damage to roots.
4. Shake plants lightly after digging to remove excess soil.

PREPARATION

1. Keep plants in an ice chest while you are in the field. If they cannot be shipped immediately, store them in a refrigerator.
2. Place plants in plastic bags without moisture, or press them between sheets of absorbent paper and encase in heavy cardboard for protection.

LABELING

Attach a label to the outside of each sample. Include the following information on labels:

1. Location where samples were taken, including names of nearby crossroads.
2. Description of specific characteristics of the site where the weeds were growing.
3. Whether plants are annuals or perennials.
4. Your name, address, and telephone number.
5. Date samples were taken.
6. Any other information that would help in the identification of the weeds.

SHIPPING

1. Contact the person or laboratory who will receive samples to determine the best method of shipping and to inform them that samples will be arriving.
2. Pack samples in a sturdy, well-insulated container to prevent crushing or heat damage.
3. Mark package clearly and request shipper to keep it in a cool location.
4. Ship packages early in the week so they will arrive before a weekend.

Weed Control Methods

Weeds and other undesirable vegetation around buildings, structures, and industrial areas can be controlled by using physical, mechanical, cultural, or chemical methods or combinations of any of these. For example, one effective technique that involves using a combination of control methods is known as *Sprinkle/Sprout/Spade-Spray* (SSSS). Using this technique, water is applied to an area that has been prepared to plant in turf or ornamentals or that is to be left unplanted. The water germinates weed seeds on the soil surface. After the seedlings emerge, they are either mechanically controlled, through hoeing or some other method, or sprayed with an herbicide. If this process is repeated a second time, up to 96% of the potential weed problem is eliminated in the top ½ inch of soil. Care must be exercised not to disturb the soil after this treatment, otherwise ungerminated seeds will be brought up to the surface and begin to sprout.

Selecting the appropriate control method or combination of methods depends on several factors. Of primary importance, as discussed above, is to be certain that the methods used are effective for the weed types, their growth habits, and their life stages. Cost of the control methods, both in dollars and in time spent, is another consideration. Some control methods may require the use of special equipment, which may or may not be available. The safety of the method to the person using it and to the public and environment is another important consideration. Sometimes control methods need to be long lasting.

Physical Control of Weeds

Physical control involves using asphalt or concrete or other barriers on top of the soil to restrict weed growth (Figure 4-7). Other barriers might include crushed rock or gravel, wood bark, plastic sheeting, or a combination of plastic with rock or bark. The use of physical barriers is usually the most effective method for controlling weeds around structures as the control is generally permanent.

Physical barriers are expensive to install and difficult to remove if no longer wanted. These methods are usually reserved for high-value structures or industrial buildings where landscaping is not practical. In these locations, physical barriers can be part of a selective landscaping plan designed for low maintenance, or the paving may serve as walkways, parking lots, or storage areas.

Sometimes weed seeds germinate in cracks and seams of physical barriers such as asphalt or concrete pavement, and growing weeds can contribute to the deterioration of these materials unless they are promptly removed or destroyed.

Mechanical Weed Control

Mechanical controls for weeds include mowing, cultivating, tilling, hoeing, and other mechanical processes that destroy the weed plants or disrupt their growth (Figure 4-8). These methods do not permanently destroy weeds and have no effect on seeds. Mechanical methods must be repeated periodically during times when weeds are sprouting or actively growing. Mechanical control methods generally work best for annual weeds and are most effective when plants are small. Perennial plants that sprout from rooting structures are not as well controlled by cultivation, because the rooting structures may be broken up into many pieces that could resprout.

FIGURE 4-7.

Materials such as asphalt, concrete, or stone provide excellent barriers to weeds.

FIGURE 4-8.

Mechanical methods are excellent ways to control some types of weeds, although they have little effect on weed seeds. However, mechanical methods may break up and disperse rooting structures of perennial plants, which may later resprout.

Cultural Control of Weeds

Cultural methods are used to alter the conditions that favor weed growth; these factors include water, nutrients, and suitable soil (Figure 4-9). The growth of weeds around structures may be decreased to some extent if water is not available, for example. This method works for summer annuals in

FIGURE 4-9.

Irrigation is a cultural method that can be used to control weeds. Irrigating an area causes weed seeds to sprout so they can be controlled by mechanical methods. During dry periods, withholding irrigation may cause emerged weeds to die.

FIGURE 4-10.

Groundcovers can be used in some areas to control weeds. Dense plantings of the groundcover compete with weeds for space, water, nutrients, and light. Some groundcovers can tolerate dry conditions.

parts of California where there is little or no summer rainfall. Water may also be applied to cause weed seeds to sprout, after which they are left to dry out or are controlled by mechanical or chemical methods.

In some areas it may be possible to control weed growth by planting groundcovers or other ornamental plants that compete with the weeds for space, light, water, and nutrients (Figure 4-10). Groundcovers can be selected that are well adapted to environmental conditions around a structure so they will require minimal maintenance.

Chemical Weed Control

Chemical control involves the application of herbicides to prevent weeds from emerging or to cause emerged weeds to die. Some herbicides destroy weeds by damaging leaf cells and causing plants to dry up. Others alter the uptake of nutrients or interfere with the plant's ability to grow normally or convert sunlight into food. The mode of action often dictates when and how an herbicide is used and what type of plant it is effective in controlling.

Preemergent herbicides kill sprouted or swollen seedlings before they emerge from the soil surface. *Postemergent* herbicides kill already emerged weeds. Some herbicides, such as diphenamid and pronamide, are generally used for preemergent control but also have some activity as postemergent herbicides.

Postemergent herbicides may destroy emerged weeds in two different ways. *Contact* herbicides usually affect only those leaves and plant parts on which they are directly deposited, therefore thorough coverage is usually essential for adequate control (Figure 4-11). Herbicides with *systemic* activity are taken up through the leaves and other green parts of the plant and are transported to roots and other growing points (Figure 4-12). These herbicides interfere with chemical processes within plant tissues to make weeds die, so it may not be necessary to have the spray cover all of the plant's foliage. Table 4-3 summarizes the mode of action of major herbicide groups.

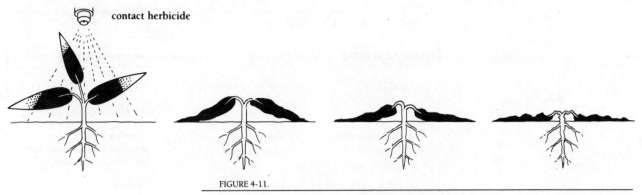

FIGURE 4-11.

Contact herbicides kill weeds by destroying leaf and stem tissues, reducing the plant's ability to transport nutrients and water or carry out photosynthesis.

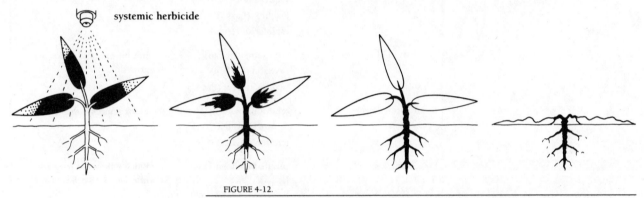

FIGURE 4-12.

Systemic herbicides are transported from leaf or stem tissues to the plant's root system. These herbicides interrupt the normal functioning of the roots, causing the plants to die.

Herbicides can be used either selectively in a crop or ornamental setting or nonselectively in a noncrop or bare ground situation. Often, the goal in controlling weeds around commercial buildings and other structures is to eliminate all vegetation, unless the area is landscaped with ornamental plants. Specific herbicide characteristics and the rates at which they are applied often influence the degree of selectivity. For example, the herbicide simazine is selective (kills only certain weed plants) when applied at low rates and nonselective (kills all plants) at higher rates. Label instructions provide information on what rates to apply for different types of control.

CHOOSING THE CORRECT HERBICIDES

Herbicides used to control weeds around structures or in industrial areas must be labeled for this purpose (Figure 4-13). Identify the weeds before you choose an herbicide, as most herbicides only control certain species. Select appropriate herbicides based on the weed species present, the life stage of the weeds, length of control desired, activation requirements (such as rainfall

TABLE 4-3

Mode of Action of Major Herbicide Groups.*

CHEMICAL TYPE	EXAMPLES	MODE OF ACTION
Inorganics	sodium chlorate	Desiccant
Petroleum oils		Physical toxicant
Organic arsenicals	MSMA, DSMA, cacodylic acid	Interfere with cellular respiration and metabolism and other functions
Phenoxyaliphatic acid	2,4-D, 2,4,5-T diclofop methyl (Hoelon)	Multiple actions
Amides	propanil (Kerb) napropamide (Devrinol) alachlor (Lasso) metolaclor (Dual)	Inhibit root and shoot growth
Substituted anilines	trifluralin (Treflan) oryzalin (Surflan) pendimethalin (Prowl)	Inhibit root and shoot growth
Substituted ureas	tebuthiuron (Spike) diuron (Karmex) fenuron (Dybar)	Block photosynthesis
Carbamates	propham (Chem-Hoe) barban Islam (Asulox)	Block photosynthesis and interfere with cell division
Thiocarbamates	molinate (Ordram) cycloate (Ro-Neet) butylate (Sutan)	Interfere with cellular respiration and metabolism, block photosynthesis, and inhibit root and shoot growth
Triazines	atrizine (Aatrex) simazine (Princep) metribuzin (Lexone) cyanazine (Bladex)	Block photosynthesis
Aliphatic acids	TCA dalapon	Unknown
Substituted benzoic acids	dicamba (Banvel) DCPA chloramben (Amiben)	Unknown
Phenol derivatives	dinoseb	Destroy cell membranes, also a desiccant
Nitriles	dichlobenil (Casoron) bromoxynil (Buctril)	Interfere with cellular respiration and metabolism and inhibit carbon dioxide fixation

*Some materials named on this table may no longer be registered for use as herbicides. Check current herbicide labels for registered uses.

CHEMICAL TYPE	EXAMPLES	MODE OF ACTION
Bipyridyliums	diquat, paraquat	Destroy cell membranes, desiccant, and block photosynthesis
Microbials	Phytophthora palmivora (Devine)	
Uracils	bromacil (Hyvar-X) terbacil (Sinbar)	Block photosynthesis
Sulfonylureas	chlorsulfuron (Glean) sulfomethuron-methyl (Oust)	Interfere with cell division
Miscellaneous	endothall glyphosate (Roundup) oxyfluorfen (Goal)	Inhibit metabolism and protein synthesis

or irrigation), and the presence or absence of crop and ornamental plants in the area. Once weeds are identified, read the herbicide label to be sure that the species of weeds you need to control are listed.

Sometimes charts showing the susceptibility of different weed species to selected preemergent and postemergent herbicides are available. These charts should be consulted before selecting an herbicide. Contact an advisor in your local University of California Cooperative Extension office for this information. Keep in mind that regional differences in climate, soil, and weed biotypes may affect the susceptibility of weeds to herbicides.

Seedlings of most weeds are quite different from mature plants, so learn to recognize both stages.

If several species of weeds are present in one location, it may be necessary to use a combination, or tank mix, of two or more herbicides to achieve

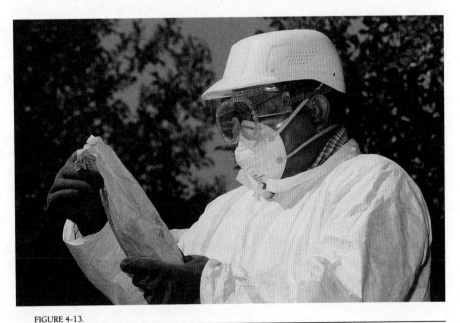

FIGURE 4-13.

Before using an herbicide, read the label to be sure the intended use is allowed.

complete control. Check the herbicide labels to be sure that the combined herbicides will be compatible and that there are no restrictions prohibiting them from being mixed together.

Learn to recognize the life stage of the weed plants being controlled, because some herbicides are not effective for all life stages. Most herbicides work best when applied to early growth stages. Older and larger weeds may require more concentrated mixtures of appropriate herbicides to achieve adequate control.

Also consider the location of the weedy site in relation to environmentally sensitive areas before selecting an herbicide. Some herbicides, for instance, are mobile in the soil and may leach to contaminate groundwater or run off to pollute surface water. Be sure you know the locations of groundwater *pesticide management zones* and endangered species before using herbicides. You must also avoid drift problems. Drift from herbicides has been known to seriously damage nontarget plants over a mile away from the application site. Some herbicides that are suitable for use around structures may be damaging to agricultural crops or landscape plants and must not be used if these are present or near by and there is any chance of the herbicide contacting them.

Soil type and organic matter content may influence the effect a soil-applied herbicide has on target weeds and may also influence the mobility of the herbicide in the soil. In general, sandy soils retain herbicides less than soils that range toward clay. Large amounts of organic matter in the soil ties up some herbicides and makes them less effective, especially if the herbicides are applied at lower label rates. Be sure to read the pesticide label to see if any adjustments in dosage must be made to compensate for soil type or organic matter content. The use of some herbicides is prohibited on certain types of soils.

APPLYING HERBICIDES

Herbicide applications around structures are usually restricted to fairly small areas. Generally, a backpack or compressed air sprayer with a 3- to 5-gallon tank is best for this purpose (Figure 4-14). These types of sprayers can be easily carried into areas where larger sprayers cannot be taken. They are inexpensive and convenient for filling and mixing, easy to use, and easy to clean.

In some situations, a powered sprayer pulled behind a vehicle may need to be used to cover a large area (Figure 4-15).

Sprayers used for herbicide application should have a fiberglass, polypropylene, or polyethylene tank as some herbicides are corrosive or chemically react with metal tanks (Figure 4-16). When using wettable powder, dry flowable, or emulsifiable concentrate formulations in backpack or compressed air sprayers, you need to thoroughly mix the pesticide before pressurizing the sprayer and shake or agitate the tank during the application to keep the suspension of herbicide in water uniform. Larger powered sprayers should be equipped with a hydraulic or mechanical agitator for this purpose. Table 4-4 is a guide for selecting herbicide application equipment.

Designate a sprayer to be used specifically for the application of herbicides and *never use this sprayer for any other type of pesticide application*. Some herbicide residues are difficult to remove from application equipment, so by designating one sprayer for herbicide use only, you will prevent possible problems of herbicide residues injuring or contaminating nontarget plants.

FIGURE 4-14.

Small backpack or compressed air sprayers are very useful for making spot-treatments of herbicides. Be sure the sprayer has a noncorrosive plastic tank. Once a sprayer of this type has been used for applying herbicides, do not use it to apply any other type of pesticide.

FIGURE 4-15.

For larger areas, a powered herbicide sprayer is very useful. These can be mounted on a truck or trailer.

FIGURE 4-16.

Spray tanks made of polyethylene or polypropylene are best when using herbicides, because some herbicides are corrosive to metal. Some herbicides even react with metal tanks to produce hydrogen gas. Plastic tanks are easily cleaned and do not absorb spray materials.

Sprayer Pressure

Apply the herbicide at low pressure and use nozzles that produce large droplets. On hand-held equipment, move the spray nozzle around so that all surfaces of emerged weeds are covered. Be sure that the spray is applied uniformly to avoid missing some of the weeds. However, confine the herbicide application to just the target plants or soil surfaces. Prevent drift onto non-target plants by spraying on windless days.

Adjuvants

Check the herbicide label to see if a spreader-sticker, surfactant, or other type of adjuvant is recommended (Table 4-5). Certain adjuvants provide better contact with leaf surfaces because they reduce the surface tension of the water in the spray mixture. Some surfactants improve the penetration of the herbicide through the cuticle of plant surfaces so that it is taken up more efficiently.

Calibration

Unlike most insecticides or fungicides, herbicide rates are based on a certain amount of the chemical applied to an acre or to an area of 1000 square feet. Even if you have selected the proper chemical and put the proper amount of it in the tank, serious problems with under- or overapplication can occur if the equipment is not properly calibrated.

To properly calibrate the herbicide applicator you are using, refer to guidelines in Volume 1 of this series, *The Safe and Effective Use of Pesticides,* or some other publication that describes calibration procedures. You may also obtain assistance from University of California Cooperative Extension advisors, equipment sales representatives, or pesticide chemical sales representatives.

TABLE 4-4

Selecting Herbicide Application Equipment.

	TYPE	USES	SUITABLE FORMULATIONS	COMMENTS
	HAND-OPERATED HERBICIDE APPLICATION EQUIPMENT (liquid)			
	Compressed air sprayers	Treating small areas, spot treating weeds.	All formulations. Wettable powders and emulsifiable concentrates require frequent shaking.	Must have plastic tank. Once used with herbicides, do not use for other purposes.
	Backpack sprayers	Same uses as compressed air sprayers.	All formulations. Wettable powders and emulsifiable concentrates require frequent shaking.	Same as compressed air sprayers.
	Wick applicators	Used for applying contact herbicides to emerged weeds.	Only water-soluble herbicides.	Simple and easy to use. Clean frequently.
	POWERED HERBICIDE APPLICATION EQUIPMENT (liquid)			
	Controlled droplet	For contact and systemic herbicides.	Usually water-soluble formulations.	Plastic parts may break if handled carelessly.
	Low-pressure sprayers	Suitable for large areas and for applying pre- and post-emergent herbicides.	All liquid formulations.	Plastic tanks are best. Should be equipped with agitator. Requires frequent maintenance. May be mounted on truck, trailer, or tractor.
	GRANULE APPLICATION EQUIPMENT			
	Hand-operated granule applicators	For broadcasting granule herbicide formulations.	Granules or pellets.	Suitable for small areas. Easy to use.
	Mechanically driven granule applicators	Turf and other landscape areas.	Granules or pellets.	Requires accurate calibration.
	Powered granule applicators	Suitable for large areas.	Granules or pellets.	Frequent servicing and cleaning is required.

Protective Clothing for Herbicide Application

When applying herbicides, wear a long-sleeved shirt and long pants, rubber boots, rubber gloves, and protective eyewear. Protective eyewear is usually *required* for herbicide applications in California, even though the label may

TABLE 4-5

Comparison of Adjuvants.

FUNCTION	Surfactant	Sticker	Spreader Sticker	Extender	Activator	Compatibility Agent	Buffer	Acidifier	Deposition Aid	Defoamer	Thickener
Reduce surface tension	■		■		■						
Improve ability to get into small cracks	■		■								
Increase uptake by target	■		■		■			■			
Improve sticking	■	■	■								
Protection against wash-off/abrasion	■	■	■	■							
Reduce sunlight degradation		■	■	■							
Increase persistence			■	■							
Reduce volatilization	■	■	■						■		■
Improve mixing						■	■	■			
Lower pH							■	■			
Slow breakdown							■	■			
Reduce drift									■	■	■
Eliminate foam										■	
Increase viscosity									■		■
Increase droplet size									■		■

not state that it must be worn. Consult the herbicide label for any other required protective clothing. Avoid walking through areas that have been sprayed.

Herbicide Tolerance

Some weeds become tolerant to herbicides that have been repeatedly used for their control. Frequent use of the same herbicide selects for individual plants of a weed species that have tolerance qualities. As these plants reproduce, the population comes to consist mostly of tolerant plants.

To slow the occurrence of herbicide tolerance, (1) reduce the frequency of herbicide applications; (2) alternate successive applications with different types of herbicides; (3) use mechanical methods to destroy weeds that survive after applying an herbicide—do this before the surviving weeds can produce seed; and (4) *always* combine the use of herbicides with other methods of control such as mulches, physical barriers, and mechanical methods.

5 Pests On or Near Food

Cockroaches, ants, and flies are some of the most common pests found on or near food in buildings. Successful suppression of these pests is based on understanding their habits so that control methods can be directed to susceptible life stages. To accomplish this, these insects must be correctly identified. General descriptions are included in the following section along with drawings or photographs of some of the more common species. For more complete information on identifying cockroaches, ants, or flies, refer to some of the identification resources listed in the References section at the end of this manual.

COCKROACHES

Several species of cockroaches inhabit buildings and may become persistent and troublesome pests. In fact, evidence of their coexistence with people throughout history is testimony to how adaptable cockroaches are to the habits of people. Buildings protect cockroaches from weather and natural enemies and contain sources of food and water as well as ample places for them to hide.

There is a difference of opinion on the classification of cockroaches. According to many experts, cockroaches belong to the insect order Orthoptera; other experts consider cockroaches and praying mantids to belong to a separate order, the Dictyoptera.

Young or immature cockroaches resemble adults (that is, they undergo gradual metamorphosis) and have similar feeding habits. Adults of many species have wings, although many species do not fly; all immature roaches, however, are wingless. Cockroaches are major pests of homes, restaurants, hospitals, warehouses, offices, and other structures that have food handling areas. These insects can contaminate food and eating utensils, destroy fabric and paper products, and impart stains and odors to surfaces they contact.

Cockroaches, especially species that live in contact with human feces like the American cockroach, may transmit bacteria responsible for food poisoning, such as *Salmonella* and *Shigella,* and viral hepatitis organisms. German cockroaches are also believed capable of transmitting staphylococcus, streptococcus, and coliform bacteria and are known to be responsible for allergy and asthma problems. In addition, German cockroaches have been implicated in the spread of typhoid, dysentery, and leprosy organisms.

Although there are more than 50 described species of cockroaches in the United States and more than 3500 worldwide, only 5 species are major pests in California (Figure 5-1). These species are the German cockroach, the oriental cockroach, the American cockroach, the brownbanded cockroach, and the smokybrown cockroach. Occasionally other species are introduced into an area and become pests in buildings. Sometimes species that usually occur outdoors invade a building.

FIGURE 5-1.

There are over 3500 species of cockroaches throughout the world. However, in California, only about 5 species are major pests.

It is important to be able to identify the species of cockroach that is causing a problem. You must also be familiar with its behavior, because each species has peculiar habits that influence the type of management methods used and locations where control efforts should be emphasized.

Cockroaches are nocturnal. They hide in dark, warm areas, especially narrow spaces where surfaces touch them on both sides. Cockroaches tend to congregate in corners and generally travel along the edges of walls or other surfaces. Periods of greatest activity differ depending on the species. For instance, brownbanded cockroaches exhibit their peak of activity in the middle of the dark cycle, while German cockroaches begin moving about within ½ hour after lights are out.

German Cockroach
Blatella germanica

German cockroaches (Figure 5-2) are the most common indoor species. They are pests in locations such as homes, hospitals, prisons, zoos, and restaurants and even become pests on ships, planes, and buses. In buildings, they are found in food preparation areas, kitchens, and bathrooms because they favor warm, humid atmospheres; areas where temperatures are around 70° to 75°F are most suitable. With severe infestations, they may occur in other parts of buildings. German cockroaches have been observed migrating in large numbers from areas of high population density to infest other locations.

Description and Seasonal Development. Adults are about ½ inch long, brown, and have two darker longitudinal bands or streaks on the top of the thorax. German cockroaches do not normally fly but are capable of gliding

GERMAN COCKROACH

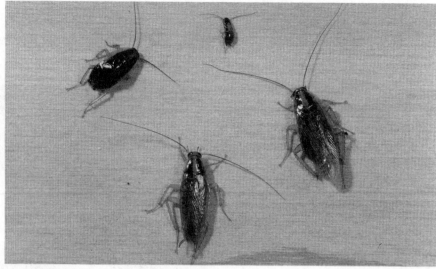

FIGURE 5-2.

German cockroach, Blatella germanica.

FIGURE 5-3.

As with many species of cockroaches, female German cockroaches lay their eggs into a capsule called an ootheca. German cockroach females carry these egg cases throughout most of the incubation period.

flight. This species has the highest reproductive potential of all the common pest cockroaches. Egg laying occurs more frequently during warm weather. Females produce about 30-50 eggs at a time, contained in a capsule called an *ootheca* (Figure 5-3) which is attached to the tip of the abdomen and carried throughout most of the incubation period; they drop the ootheca about one day before egg hatch. Eggs of this species hatch in about 28 days at room temperature, and the hatchlings reach maturity in about 40 to 125 days after passing through five to seven molts. Adult females live about 200 days, producing six to eight oothecae throughout this time.

Oriental Cockroach
Blatta orientalis

The oriental cockroach (Figure 5-4) lives in colonies in dark, damp places. Usual places where they are found include indoor and outdoor drains, water control boxes, woodpiles, basements, garbage chutes, damp areas

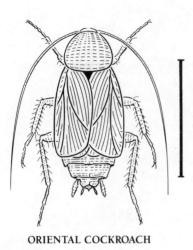

ORIENTAL COCKROACH

FIGURE 5-4.

Oriental cockroach, Blatta orientalis.

under houses, and trash cans. They may be found in large numbers outside where people feed their pets. At night, oriental cockroaches may migrate into buildings in search of food. Migrations may also be observed during periods of adverse weather. They usually remain on the ground floor of buildings and move more slowly than other species. They do not fly and are unable to climb smooth surfaces, consequently they are commonly found trapped in porcelain tubs or sinks.

Description and Seasonal Development. The adult is about 1 to 1¼ inches long and dark brown, almost black. Males have fully developed wings which are shorter than the body, but are flightless; females have rudimentary wings. Females deposit an average of eight egg capsules (oothecae) during their lifetime; each capsule produces about 16 young. Oothecae are carried by the female for about a day after formation, then deposited in a humid location where the eggs complete development. Eggs require about 60 days to hatch; the time for development from egg hatch to adult ranges between 300 and 800 days, depending on environmental conditions. After reaching maturity, adult females live for another 30 to 180 days.

American Cockroach
Periplaneta americana

The American cockroach (Figure 5-5) is a common pest in zoos and animal rearing facilities and is found in sewers, water meter boxes, and steam tunnels. They prefer very warm and humid environments (temperatures in excess of 82°F) but occasionally forage from sewers and other areas into buildings. Their foraging is confined mostly to the ground floor of buildings unless suitable conditions exist in higher locations.

Description and Seasonal Development. The American cockroach is one of the largest cockroaches that invade buildings. Adults are about 1½ to 2 inches long and are reddish brown. Females carry oothecae for a short period before depositing them in a safe location. Females produce between 10 to 60 oothecae, each giving rise to about 14 young. Eggs require about 45 days to incubate. Young mature in as little as 215 days; some studies show that maturation takes up to 400 days. The average life span for an adult female is about 440 days.

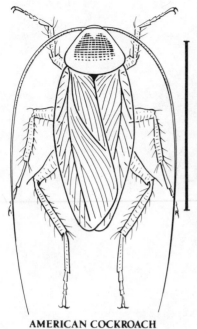

AMERICAN COCKROACH

FIGURE 5-5.

American cockroach, Periplaneta americana.

The greatest numbers of adults are usually seen during late summer months.
American cockroaches can fly and are attracted to street lights at night.
They are active throughout the year in temperatures of 70°F or higher.

Brownbanded Cockroach
Supella longipalpa

The brownbanded cockroach (Figure 5-6) is the second most common
indoor cockroach species in California. Individuals may be widely distributed
throughout a building, hiding behind pictures, beneath furniture, among books,
and in other drier areas not normally infested by German cockroaches. They
seek out areas that are warm most of the time such as radios, televisions, and
refrigerators. They appear to be more common in apartments or homes of

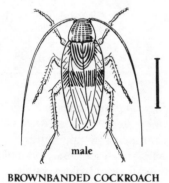

male

BROWNBANDED COCKROACH

FIGURE 5-6.

Brownbanded cockroach,
Supella longipalpa.

FIGURE 5-7.

*The banding pattern on brownbanded cockroaches varies as the individual cock-
roach matures.*

the elderly who don't use air conditioning. They are also seen in animal rearing facilities and are becoming more frequent pests in institutional kitchens, offices, and hospitals.

Description and Seasonal Development. Brownbanded cockroaches are about ½ inch long at maturity. Adult males are golden brown and have a narrow body; their wings extend beyond the tip of the abdomen. Female adults are darker chestnut brown, have a teardrop-shaped body, and their wings do not completely cover the abdomen. Both sexes have distinctive horizontal bands of color as shown in Figure 5-7. Nymphs have two pale bands which run horizontally across the body. Wings have brownish yellow stripes; adult males fly readily when disturbed but females do not fly. Females carry their oothecae for 24 to 36 hours before attaching them to a hidden vertical surface. They often glue their eggs in clusters on furniture or in appliances. Eggs require about 70 days for incubation, and about 160 days are required to reach maturity. More eggs are produced during summer months. Brownbanded cockroaches prefer warmer temperatures (greater than 80°F) than the German cockroach, therefore these two species are rarely found together.

Smokybrown Cockroach
Periplaneta fuliginosa

The smokybrown cockroach (Figure 5-8) is usually found in garages, decorative plantings and planter boxes, woodpiles, and water meter boxes; it also occasionally lives in municipal sewers. This species is very visible and is usually seen by building occupants. It can be found in upper parts of buildings, including attics, although it is not specifically attracted to heated areas of buildings where other species are likely to be found; it also crawls under shingles or siding and can be found in trees, shrubs, and other vegetation during summer months.

Description and Seasonal Development. The adult is about 1¼ inches long and uniformly dark brown to mahogany, sometimes almost black. This species has well-developed wings, ordinarily flies, and is known to be attracted to

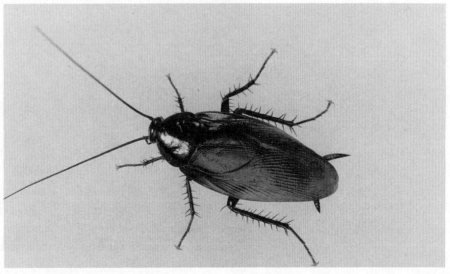

FIGURE 5-8.

Smokybrown cockroach, Periplaneta fuliginosa.

lights at night. The female carries its ootheca for about one day before attaching it to a surface. Eggs hatch on an average of 45 days after being laid and about 20 young emerge from each ootheca. Females reach maturity in about 320 days after hatching. More adults occur during summer than during other season.

Management Guidelines for Cockroaches

Although the German cockroach is most common, it may be possible for a building to be inhabited by more than one cockroach species. Successful management depends on identifying the species involved and then selecting methods of control that are effective against these species. You must also change the conditions that attracted and favored the infestation in the first place.

Carefully inspect the infested area to locate cockroaches. Nighttime surveys are useful as cockroaches are nocturnal (active at night). Use a flashlight and search in cracks, under counters, around water heaters, and in other dark locations (Figure 5-9). Look for live and dead cockroaches, cast skins, egg capsules, and droppings, all of which aid in identification. Use sticky traps or jar traps (Figure 5-10) to monitor cockroach activity and capture specimens for identification; place traps along walls and other areas where cockroaches are known to travel. Traps must be placed right next to walls or other objects or in intersections for maximum effectiveness. If a thorough inspection fails to produce results, a pyrethrin flushing agent may help to dislodge hiding roaches so they can be captured and identified. Synergized pyrethrins cause

FIGURE 5-9.

A flashlight is essential to search for cockroaches in cracks, under counters, and in other dark locations.

FIGURE 5-10.

Sticky traps are helpful for monitoring cockroach populations. These are most effective if an attractant is placed in the trap.

rapid paralysis of cockroaches. However, although extremely useful as a flushing agent, pyrethrin is less effective for the control of cockroaches than other insecticides because it does not have residual action; some cockroaches may recover from the toxic effects of a pyrethrin knockdown spray.

Once the species is identified, begin to plan your control strategy. Map locations of suspected or actual infestation and concentrate control measures on these areas. Your management program should include nonchemical methods such as sanitation and exclusion whenever possible, because chemical control applied without attention to sanitation may not always be successful.

Sanitation. Eliminate sources of food and water wherever possible. Food should be stored in roachproof containers such as glass jars or sealable plastic dishes. Keep garbage and trash in containers with tight-fitting lids. Remove trash, newspapers, rags, boxes, and other items that provide hiding places and harborage. Eliminate plumbing leaks and correct other sources of free moisture. Increase ventilation where condensation is a problem. Vacuum all cracks and crevices to remove debris and food. Be sure surfaces where food or beverage spills have taken place are cleaned up immediately. Trim shrubbery around buildings to increase light and air circulation, especially near vents. Remove trash and stored items around the outside of buildings that provide hiding places for cockroaches.

False-bottom cupboards, hollow walls, and similar areas are common refuges for cockroaches. In locations where these conditions exist, and where cockroach populations are high and difficult to control, remodeling may be warranted to eliminate infestations. Commercial areas where food is prepared, stored, or sold would most benefit from remodeling to eliminate cockroach hiding places. If remodeling is too costly or impractical, treat these areas with inorganic dusts such as silica gel or boric acid powder. Boric acid powder has been shown to be less repellent to cockroaches than silica gel.

Exclusion. If roaches are migrating into a building from outdoors, seal cracks and other openings to the outside. Look for other methods of entry, such as from items being brought into the building. Look for ootheca glued to undersides of furniture, in refrigerator and other appliance motors, boxes, and other items. Locate cracks inside the treatment area where cockroaches can hide; seal these with caulking.

Chemical Control. Insecticides are usually very effective in controlling cockroaches. However, when infestations occur in food storage, preparation, or serving areas, it is important that insecticides be used with extreme care. Use only materials that are registered for use in food preparation areas. Combine chemical controls with other nonchemical methods whenever possible. In all areas, use only registered insecticides and follow the label directions carefully.

If contact activity or rapid control is needed, use a quick-acting, low-residual insecticide such as a synergized pyrethrin formulation or a synthetic pyrethroid. These insecticides are usually applied as a liquid spray covering general areas or as a mist in enclosed areas.

To control roaches that will be hatching from eggs, have been missed by other treatments, or are migrating in from outside the treatment area, use an insecticide with residual activity. Residual materials may be formulated as liquid sprays, dusts, desiccants (inert dusts or sorptive powders), or impregnates such as lacquers and paints. One possibility for residual control is the use of an insect growth regulator (IGR), which sterilizes or kills immature nymphs. Although highly effective, it may take many months to control a cockroach population with an IGR. For this reason, it is common practice to combine an IGR with an insecticide having more rapid action to provide an effective long-term control program.

Residual insecticides may be formulated as dusts, liquid sprays, paints, or baited traps. Dusts may have desiccant action, such as silica gel, or be internal or contact poisons, such as boric acid. Blow dusts into cracks and crevices or lightly spread them in areas where visible residues are not a problem and where people will not contact them. Do not use boric acid dust in areas subject to repeated moisture followed by drying as the dust will cake and lose its effectiveness. Boric acid is effective but slow acting, killing the pests several days after ingestion or contact. Cockroaches pick up this dust on their bodies as they walk through a treated area, then ingest small amounts when they groom themselves.

Apply liquid pesticides as sprays to cracks and crevices where roaches spend most of their time and to wall and floor surfaces and other areas where cockroaches travel; use encapsulated formulations for slow release of the active ingredient. Lacquer paints containing residual insecticides are sometimes applied to wall surfaces in areas of cockroach infestation to provide long-term, slow release of the insecticide.

Baited traps or bait stations are manufactured as plastic or cardboard units that contain an attractant. Cockroaches enter the bait station or trap through small openings. Traps and bait stations have the advantage that insecticides can be confined to a small area rather than being dispersed.

Cockroaches may avoid certain deposits of residual insecticides. For this reason, it is important to use materials that do not repel them, otherwise you must have thorough coverage to ensure that the cockroaches will contact treated areas. Cockroach populations may develop resistance to the insecticides you use for their control. Populations of cockroaches migrating in from another area may already be resistant to insecticides that were used against them elsewhere. Methods that may help to reduce resistance problems include (1) use of alternate, nonchemical control methods, such as biological control and good sanitation, (2) lowering the frequency of insecticide application, (3) alternating the types of active ingredient and formulation, and (4) using insecticides that do not repel cockroaches. Sometimes cross-resistance develops in cockroach populations. This is a condition where the resistance to one type or class of insecticide makes the insect resistant to one or more other types or classes of insecticides.

Use of Boric Acid as an Insecticide for Cockroaches

Boric acid is an effective insecticide for the control of cockroaches in buildings and does not appear to present serious health hazards to building occupants, in part because it is a nonorganic chemical which does not penetrate the skin. Boric acid powder is slow acting and may take 7 days or more to begin having a significant effect on a cockroach population. Any commercial use of boric acid as an insecticide must be as a registered pesticide, and the use must be strictly according to the label. Formulations usually contain about 1% of an additive which prevents caking and improves dusting properties. Boric acid does not decompose and therefore retains its effectiveness as long as it does not become wet.

Where to Use: Dust boric acid powder into out-of-the-way places where cockroaches hide. Remove the kick panels from the front of appliances and apply boric acid powder to the entire area underneath. Drill ½ inch holes at the top of kick panels beneath cabinets and blow liberal amounts of dust into these areas. Dust spaces under sinks and in dead spaces between sinks and walls. Also dust areas where utility pipes pass through walls. Sprinkle dusts in intersections and in corners of shelves in cupboards.

Method of Application: Boric acid powder can be applied with a bulb duster, hand-operated or power blower, or sprinkled from a small container. Be sure the application method is consistent with label instructions. Avoid dispersing the powder into air ducts or food, eating utensils, food preparation surfaces, and other locations where people or pets might come in contact with it.

Application Rates: Follow label information for application rates. An apartment will usually require 1 to 2 pounds of material; an average size house may need 2 to 4 pounds. Commercial establishments such as restaurants, cafeterias, hotels, and hospitals may require 30 to 40 pounds.

Precautions: Wear gloves and goggles while applying boric acid powder to avoid the powder contacting your skin or eyes. Do not inhale the dust. This material is highly irritating to the eyes and respiratory system. Avoid application in areas where children or pets may come in contact with the powder. Do not apply to plants or to soil where plants are growing.

Monitor and Evaluate. After a cockroach control program has been started, evaluate the effectiveness of the methods that are being used. Use traps or visual inspections to help determine if further treatment is necessary. If populations persist, reevaluate the situation. Look for other sources of infestations, make sure that all possible entryways are blocked, be certain that food and water sources are eliminated as much as possible, and continue sealing and eliminating hiding places. Repeat insecticide applications if necessary. However, if insecticides appear to be less effective, resistance may be occurring. Overuse of insecticides and indiscriminate application may cause resistance problems.

If cockroach populations are controlled, continue monitoring with baited traps on a regular basis to make sure reinfestation is not taking place. Maintain sanitation and exclusion techniques to avoid encouraging a new infestation. If severe reinfestation continues to occur, consider having the infested areas modified or remodeled to reduce the amount of suitable habitat for cockroaches.

ANTS

Ants are among the most prevalent pests in households. They are also found in restaurants, hospitals, offices, warehouses, and other buildings where they can find food and water. Most ants can bite with their pincerlike jaws (few actually do), and some have venomous stings. However, they are annoying pests primarily because they appear in large numbers in buildings and may nest in wall voids or other parts of structures (Figure 5-11). Ants contaminate and destroy some agricultural products and stored foods. Certain species stain or cause feeding damage to textiles. On outdoor plants, ants protect and care for honeydew-producing insects (aphids, soft scales, and mealybugs), which may interfere with the natural biological control of these pests. In nature, ants may perform beneficial functions by preying on certain species of insect pests and aerating soils.

Throughout the world there are over 12,000 species of ants. In California, less than a dozen species are important pests in buildings, and a similar number cause problems in agricultural and landscape areas.

Ants belong to the insect order Hymenoptera and are close relatives of bees and wasps. These insects undergo complete metamorphosis, passing through egg, larval, pupal, and adult stages. Larvae are immobile and wormlike and do not resemble adults. Ants, like many other hymenopterans, are social insects with duties divided among different types, or castes, of adult individuals. Queens conduct the reproductive functions of a colony; they lay eggs and participate in the feeding and grooming of larvae. Sterile female workers gather food, feed and care for the larvae, build tunnels, and defend the colony. Males do not participate in colony activities; their only apparent purpose is to mate with the queens. Few in number, males are fed and cared for by the workers.

FIGURE 5-11.

Ants are among the most prevalent household pests; they are also found in restaurants, hospitals, offices, warehouses, and other buildings.

Ants that invade homes and buildings include the Argentine ant, odorous house ant, Pharaoh ant, thief ant, and several species of carpenter ants. Other species such as fire ants may be occasional indoor pests in parts of California. Harvester ants, velvety tree ants, and other species nest outdoors but also occasionally invade structures. The common building pests are described in this chapter except for carpenter ants, which are discussed in Chapter 9, "Wood-Destroying Pests."

Argentine Ant
Iridomyrmex humilis

The Argentine ant (Figure 5-12) is the most common household and building pest in central and southern California. This species forages in restaurants, grocery stores, offices, schools, warehouses, and any other location where suitable food and water are available. It is a persistent pest which is difficult to control once it has established a colony inside or near a building. This ant is not a native species, but was introduced into the United States around 1890.

Description, Development, and Habits. The adult worker is about ⅛ inch long and is light to dark brown. Queens are lighter colored and are from ³⁄₁₆ to ¼ inch long; several hundred queens may live in a single, large colony. Argentine ants usually nest in the soil; they are often found next to buildings or along sidewalks. They also construct nests under boards and plants and sometimes under buildings. They occasionally make nests in wall voids or in soil of houseplants if conditions are satisfactory. When foraging, thousands of workers form long trails from the nest to the food location. Ants can be seen traveling in both directions along these trails. Workers all share food with each other and the colony's queens.

Tiny white eggs are laid by queens throughout the year; the maximum production, between 20 and 30 eggs per day per queen, occurs during warm months. The average incubation period is 28 days under favorable weather conditions. After hatching, larvae remain in the nest and are fed, groomed, and protected by adult workers. The larval stage lasts approximately 31 days and pupation takes about 2 weeks. During warm weather, colonies usually break up into smaller groups and migrate closer to food supplies. In the winter they again aggregate into larger colonies. Mating most often takes place in the nest rather than on a mating flight. Individual queens live as long as 15 years.

Inside a building, the Argentine ant feeds on sugars, syrups, honey, fruit juice, and meat. Outdoors it is attracted to the sweet, sticky secretions called honeydew produced by soft scales and aphids; it also feeds on dead insects and other arthropods and decomposing tissues from dead animals. This diversified diet aids colony survival and success because there is usually always some type of food available. The Argentine ant's high reproductive potential (a result of the large number of queens in each colony) and the ability of a colony to rapidly adapt and settle into nest sites in a great variety of buildings and natural locations also contribute to this species' success. New colonies can be set up quickly and grow rapidly in size because queens mate in the nest and participate in the feeding and grooming of larvae. The Argentine ant has no important natural enemies in California.

Management Guidelines for Argentine Ants. Argentine ants enter buildings seeking food and water, warmth and shelter, or a refuge from dry, hot weather. They may appear suddenly in buildings if other food sources become unavailable.

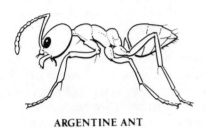

ARGENTINE ANT

FIGURE 5-12.

Argentine ant,
Iridomyrmex humilis.

For example, rainfall may wash away honeydew from nearby aphid- or scale-infested plants, forcing ant workers to search out a new food source.

Ant management requires diligent efforts and the combined use of mechanical, cultural, sanitation, and chemical methods of control. Emphasis should be on excluding ants from buildings and eliminating food and water sources.

To keep Argentine ants out of buildings, caulk cracks and crevices around foundations that provide entry from outside. Trees and shrubs located near buildings may attract Argentine ants if they have aphid or soft scale infestations and foliage is coated with honeydew. Honeydew is an attractive food for ants; ants protect colonies of aphids or soft scales from natural enemies and may even move the insects to other plants to maintain the honeydew production (Figure 5-13). Plants such as bamboo, cherry laurel, and fig trees, which are especially attractive to Argentine ants because of their aphid or soft scale populations, should not be planted near buildings. Develop a program to control aphids and scale insects if infested trees or shrubs cannot be removed. A sticky resin or petrolatum can be applied in a band around the lower part of the trunk of a plant to provide a barrier to the ants (Figure 5-14). When ants are excluded, natural enemies of aphids or scales are more effective.

Indoors, eliminate cracks and crevices wherever possible, especially in kitchens and other food preparation and storage areas. Attractive food items such as sugar, syrup, honey, and other sweets should be stored in closed containers

FIGURE 5-13.

Honeydew, produced by plant scale insects, is an attractive food for ants.

FIGURE 5-14.

A sticky paste works as a barrier to keep ants from entering an area or from reaching honeydew in plants infested with aphids or scale insects.

FIGURE 5-15.

Bait used for the control of ants should contain a slow-acting pesticide so workers will carry it back to the nest. This will provide control of larvae as well as adults.

that have been washed to remove residues from outer surfaces. Rinse out empty soft drink containers or remove them from the building. Thoroughly clean up grease and spills. Do not store garbage indoors. Look for indoor nesting sites, such as potted plants; if ants are found, remove the containers from the building, then treat the soil with an insecticide or submerge the pots for 20 minutes in standing water that contains a few drops of liquid soap.

An insecticide labeled for ant control can provide immediate knockdown of foraging ants if necessary while sanitation and exclusion measures are being taken. However, if ants can be thoroughly washed away and excluded from an area, an insecticide is probably not necessary. Soapy water, as an alternative to insecticides, may be effective in controlling foraging ants in a building.

Chemical control of ants is most successful when the insecticide treatment is focused on queens and larvae inside nests; killing workers does little to control the colony because as few as 1% of the workers are able to provide sufficient food for nestbound queens and larvae. Follow the insecticide label carefully. Insecticides used indoors usually do not provide long-term control, so combine their use with other practices whenever possible. Outdoors, spray around foundations with a material having a long residual to create a chemical ant barrier. Apply spray to sidewalk or pavement cracks around the building perimeter. If structures are on a raised foundation, treat crawl spaces with dust formulations to provide long-term control.

Baits are the most effective way to get poison into the nest (Figure 5-15). Bait can be used most effectively when ant colonies are dispersing in search of food. For instance, set out bait in the early spring, before flowering plants begin to bloom, as ants will begin feeding on nectar as soon as it is available. In the fall, use bait when other food supplies become scarce. Have bait available after rains when honeydew is washed off plants. Place bait next to nests whenever possible, but avoid placing poisonous bait in areas where it can be found by small children. Bait is most successful if it is not competing with other food sources in or around buildings, so combine baiting with sanitation practices.

See Table 5-1 for a summary of Argentine ant treatment methods.

TABLE 5-1

Treatment Methods for Argentine Ants. Control methods for Argentine ants may vary, depending on their location.

LOCATION	TREATMENT
Nests outside building.	Use baits (keep these away from children) or apply a residual spray to nest areas and along foraging trails.
Foraging ants inside or near food preparation areas.	Apply a quick-acting, short-residual spray or wash up ants with soapy water or remove them with a vacuum.
Ants foraging for honeydew on trees or shrubs near buildings.	Band trunks of trees with sticky material to keep ants from getting to scale or aphids; wash foliage with mild soapy water to remove honeydew and help kill the plant-feeding insects.
Ants in greenhouses or animal facilities.	Use sticky material or a water or oil barrier to keep ants from climbing legs of tables or equipment.

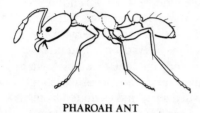

PHAROAH ANT

FIGURE 5-16.

Pharaoh ant,
Monomorium pharaonis.

PHAROAH ANT   THIEF ANT

FIGURE 5-17.

*The Pharaoh ant can be distin-
guished from the thief ant by the
three segments in its antennal
club.*

Pharaoh Ant
Monomorium pharaonis

The Pharaoh ant (Figure 5-16) is one of the most difficult ant pests to control. Although it occurs in all types of buildings, its presence in hospitals is particularly a problem because it may infest open wounds and is known to carry *Streptococcus, Pseudomonas,* and *Staphylococcus* bacteria on its body. The Pharaoh ant has had limited distribution in California, but its populations are expanding; it is principally found in several of the state's larger metropolitan areas.

Description, Development, and Habits. Workers are small, approximately 1/16 inch long, and range in color from yellowish to light brown to reddish. They can be distinguished from the similar-appearing thief ant by the three segments in its antennal club; the thief ant has two (Figure 5-17). Use a hand lens to examine the ant's antennae to assist in your identification. Queens produce 4 to 12 eggs per day throughout the year, and are most productive during warmer seasons. Under average weather conditions, eggs hatch after 7 days. The larval stage lasts 18 days, followed by a 3-day prepupal period and a 9-day pupation. Queens live approximately 39 weeks. Their diet includes sweets and fatty foods, although fats are preferred. Workers also feed on live and dead insects. Food is carried back to the nest for queens and larvae.

Pharaoh ants enter buildings from outside through cracks and other openings. They can also be transported into buildings on packages, supplies, and furnishings. These ants nest in buildings, preferring warm places such as around hot water pipes and heating systems. They colonize in wall voids, behind baseboards, between layers of flooring, and under furniture. They also build nests in odd places such as among linens, in appliances, and among paper products. Pharaoh ants travel great distances from nest sites to food and do not always follow the same path; this makes it difficult to locate their nests.

Management Guidelines for Pharaoh Ants. Adequate management requires that you exterminate Pharaoh ant colonies from inside buildings. However, this can be difficult, time consuming, and expensive. Large buildings, such as hospitals, contain many areas where colonies can exist, and many of these locations are inaccessible to pest control efforts. Sometimes control methods such as applying sprays may appear to be working, when actually they are only fragmenting the colonies into smaller, more dispersed groups. As long as food is available, smaller colonies will continue to grow and persist, and the infestation may become worse after treatment due to this dispersion.

Before attempting to control an infestation of Pharaoh ants, study the building carefully to learn where colonies may be. A few dabs of peanut butter as an attractant can aid in identifying colonies; however, the great distances between colonies and food sources may make this task difficult or at times impossible. Look for nests in linen closets, in electrical outlets, in and around appliances, near water pipes, in heating ducts, under furniture, and in drawers and cabinets.

Use sanitation practices to restrict access to food and water. Because of their small size, Pharaoh ants can get into many small places, including lidded jars lacking rubber seals. Spills must be thoroughly cleaned up, and garbage, soiled materials, and other items that serve as food should be removed from the building. If possible, seal off cracks and crevices to restrict the colony's travel to food or water.

Treat the perimeter of the building with a persistent insecticide to control ants foraging outside for food. This will also help to stop reinfestation. Be

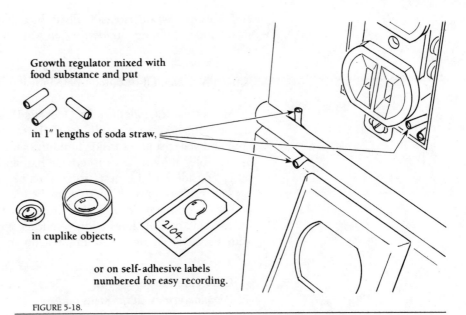

Growth regulator mixed with
food substance and put

in 1″ lengths of soda straw,

in cuplike objects,

or on self-adhesive labels
numbered for easy recording.

FIGURE 5-18.

*Bait stations for controlling ants can be simple devices such as short pieces of
soda straw.*

sure to look for trees, shrubs, wiring, or other items that provide bridges to
the building.

The principal chemical control method for Pharaoh ants inside a building involves the use of a growth regulator or slow-acting insecticide combined
with bait that can be taken back to the nest to be fed to queens and larvae.
Locate bait stations (Figure 5-18) near nests, next to ant trails, under appliances, around trash storage areas, inside cabinets, inside wall-mounted electrical outlets, and in other areas where ants are seen or suspected. Keep all
bait out of reach of pets and children. Once bait application areas are established, avoid moving or disturbing them so as not to disrupt feeding patterns.
Within one week after placing bait, check and refill bait containers if necessary. Continue rechecking and replacing bait for up to 6 weeks or until no
further activity is observed. After destroying colonies with a baiting program,
prevent reinfestation by applying a residual insecticide in indoor areas where
ants gain access to food or water. Use an insecticide in this manner only if
entryways cannot be blocked; follow the insecticide label carefully.

Odorous House Ant
Tapinoma sessile

The odorous house ant (Figure 5-19) gets its name from the foul,
musty odor emitted when an individual is crushed. Like other ant pests, the
odorous house ant invades houses and other buildings in great numbers.

Description, Development, and Habits. Workers are approximately ⅛ inch
long, slightly longer and broader than Argentine ants, and have a dark, uniform brown or black color. They forage from their nests in long trails, but are
slightly slower moving than Argentine ants. Food consists primarily of sugars; the honeydew of aphids and scale insects is their natural food source.
Numerous queens are found in each colony, although colonies do not join
together. Mating usually takes place outside of the colony. Nests are made outdoors in sandy soils, pastures, wooded areas, under stones and logs, in trees

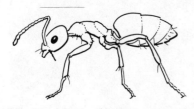

ODOROUS HOUSE ANT

FIGURE 5-19.

Odorous house ant,
Tapinoma sessile.

and tree stumps, and occasionally in bird and mammal nests. Nests are also made under building foundations, in wall voids, or around water pipes and water heaters.

In locations where the Argentine ant is also a pest, the odorous house ant is usually driven off or outcompeted for food.

Management Guidelines for Odorous House Ants. Control methods for the odorous house ant are similar to those used for the Argentine ant. Remove food and water sources inside buildings and eliminate as many entry points as possible. Indoors, use insecticide bait and spot-treatments along foraging trails. Outside, look for nests around the perimeter of infested buildings and apply a residual spray or bait to them. Reduce populations by controlling aphids and scales on plants located near buildings. Remove plants that are susceptible to aphid or scale infestation whenever possible, otherwise band trunks with a sticky resin or petrolatum to keep ants away from the honeydew.

California Harvester Ant
Pogonomyrmex californicus

The California harvester ant (Figure 5-20) is most commonly found in parking lots, sidewalks, lawns, and landscaped areas. It is an occasional invader of buildings. Individuals of this species are capable of inflicting a painful sting and are aggressive and usually attack when disturbed.

Description and Habits. The worker is about $3/16$ to $1/4$ inch long and light rusty red with lighter-colored legs: males are black and red. Nest openings are characterized by a fan-shaped mound, but its entrance is completely closed during winter months (Figure 5-21). With the onset of warm weather, nests are opened during the day, but always closed at night. California harvester ants forage during the day for small seeds and grains. They readily sting people or animals coming too close.

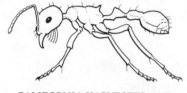

CALIFORNIA HARVESTER ANT

FIGURE 5-20.

California harvester ant,
Pogonomyrmex californicus.

FIGURE 5-21.

Harvester ant nest openings are characterized by a fan-shaped mound.

Management Guidelines for the California Harvester Ant. Control of the California harvester ant requires locating nests and physically destroying them or applying insecticides to openings. Destroy nests by digging them up with a shovel or by using a mechanical cultivator. Once nests have been opened and exposed, thoroughly saturate the area with soapy water to stop survivors from rebuilding. Removing food supplies is nearly impossible unless all seed-producing plants are eliminated in the area where the colony forages. When working around ant nests, take precautions to avoid being stung by workers. California harvester ant venom usually causes a painful reaction which has been known to persist for over 30 days.

Thief Ant
Solenopsis molesta

Thief ants (Figure 5-22) are pests in homes and other buildings. Their name is derived from the habit of building nests near colonies of other ant species and stealing food from them. They also kill and eat larvae of larger ants. These are small, light-colored ants which may be difficult to see, especially in some parts of buildings or poorly lighted areas. They are attracted to greasy foods, cheese, and animal feces. Because of their size, they can get into almost any type of container where food is stored. Their omnivorous feeding habits make it possible for them to transmit disease organisms to food items. They are persistent and difficult to control once they have invaded a building.

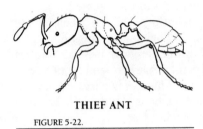

THIEF ANT

FIGURE 5-22.

Thief ant, Solenopsis molesta.

Description, Development, and Habits. Thief ants are the smallest of the ant pests to invade buildings. Workers are about $1/16$ inch long and yellowish. This species resembles Pharaoh ants, but individuals are smaller; they have a two-segmented antennal club whereas the Pharaoh ant's antennal club is three-segmented (see Figure 5-17). Workers are capable of inflicting stings if they are disturbed, although they are not very aggressive.

Eggs of the thief ant incubate for about 22 days. The larval stage lasts 21 or more days, the prepupal period 2 to 11 days, and pupation 20 days. When thief ants nest outside, their colonies are usually found under rocks or boards near nests of other ant species. They build small tunnels into the nests of larger ants, providing them access to food as well as the larvae of their host. Thief ants may also build nests in wall voids and other secure locations inside buildings.

Management Guidelines for Thief Ants. Because thief ants tend to nest outside, begin management by blocking their access into buildings. Seal cracks and openings around the perimeter of the building. Where openings cannot be sealed, use a sticky resin or petrolatum as a barrier to keep ants from entering. Specific long-residual insecticides serve as chemical barriers when sprayed around foundations or other structures that ants must cross to enter a building. Search out nests under stones, boards, and similar objects located near buildings and physically destroy them or, if destruction is not possible, treat them with a registered insecticide according to label instructions.

Remove attractive food sources or seal them in antproof containers. Thief ants, however, are capable of finding food almost anywhere, so sanitation alone will not control this particular pest. If thief ant colonies are nesting inside a building, use a poison bait that will be carried back to the nest and fed to queens and larvae. Sugar-based bait is not very effective with thief ants because they prefer greasy foods, meats, cheeses, and similar animal products.

FLIES

Flies belong to the insect order Diptera and are related to mosquitoes and gnats. These insects undergo complete metamorphosis. Fly larvae, called maggots, have a wide range of feeding habits depending on the species. Some larvae are plant-feeders and can be serious agricultural pests. Others feed on rotting or decaying remains of plants or animals, or on animal excrement. Maggots of many species are internal parasites of arthropods or vertebrates. Most adult flies are winged and fly readily. Flies and all other dipterans have one pair of wings, as opposed to other orders of winged insects which have two pairs. In place of the second pair, dipterans have knobbed balance organs called *halteres* (Figure 5-23).

Flies are serious pests in and around buildings because they can transmit disease organisms and filth. They also leave deposits of regurgitated food and excrement on walls, furniture, draperies, paintings, and other belongings. They are annoying due to their flying, buzzing, and habit of landing on food, walls, windows, and furnishings. In general they interfere with people's comfort. Flies are also pests in outdoor eating areas, open-air markets, and home yards.

Although there are several thousand species of flies, only a few are persistent pests in or around buildings. These include the house fly, the little house fly, the cluster fly, vinegar or fruit flies, and blow flies.

House Fly
Musca domestica

Little House Fly
Fannia canicularis

The adult stages of the house fly and little house fly (Figure 5-24) are typically found in buildings such as restaurants, homes, offices, hospitals, and grocery stores. They seek out these locations for shelter, food, and suitable breeding sites. Besides their annoyance, these two species are capable of carrying disease-causing organisms on their bodies. Flies can contaminate food, eating utensils, and other items with these organisms and transmit salmonella, dysentery, typhoid fever, cholera, anthrax, leprosy, yaws, and infectious hepatitis. In addition to disease organisms, these flies at times carry eggs of pinworms, whipworms, hookworms, *Ascaris,* and tapeworms, which may occasionally infect people. House flies are claimed by some to be the greatest threat

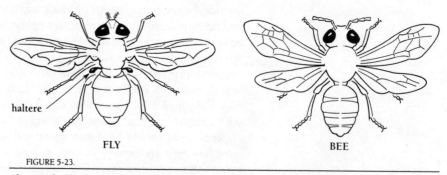

FIGURE 5-23.

Flies and all other dipterans have one pair of wings while all other winged insects have two pairs. In place of the second pair, flies have knobbed balance organs called halteres.

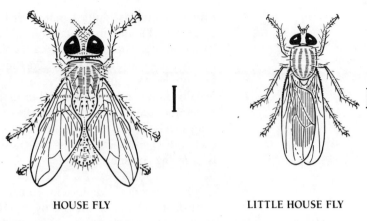

HOUSE FLY   **LITTLE HOUSE FLY**

FIGURE 5-24.

House fly, Musca domestica *(left), and little house fly,* Fannia canicularis *(right).*

to people's health of any species of insect due to their ability to transmit so many disease and parasite organisms.

House Fly Description, Development, and Habits. The adult house fly is about ¼ inch long and dark gray with four darker longitudinal stripes on the top of the thorax. The abdomen may be gray or yellowish with a darker median line and an irregular pale yellowish spot on each side near the thorax. When at rest, the wings of the house fly fold back and are held slightly away from the body but do not overlap. Larvae, or maggots, of the house fly are cream-colored, pointed at the front end and blunt at the rear end. Females lay eggs in bunches of 75 to 100 in moist animal manure or garbage; each female may, under ideal conditions, lay up to 1000 eggs. Eggs usually hatch in 8 to 12 hours. Larvae feed for 3 to 5 days, then enter a prepupal stage for an average of 2 to 4 days. Pupation usually takes 4 to 5 days before adults emerge. In total, it requires from 7 to 45 days from egg laying to adult emergence under a normal temperature range of 60° to 95°F. Below 55°F, development usually stops. During warm weather, two to three generations per month are typical.

Adult house flies are present throughout the year but are most abundant in late summer and early fall. An adult house fly lives from 30 to 60 days and can fly as far as 20 miles, but usually confines its activities within a 1 to 4 mile range; it remains within the confines of a small area if food is plentiful. During the adult stage, house flies feed on many substances including feces, decaying organic matter, and a variety of liquid foods. They are attracted to indoor food preparation and serving areas. House flies leave straw-colored spots of regurgitated food and dark spots of fecal matter on surfaces where they feed or rest.

Little House Fly Description, Development, and Habits. The little house fly is a smaller and more slender species than the common house fly. It is dull gray with three darker longitudinal stripes on the top of the thorax. When at rest, the wings of the little house fly are partly folded over each other and are held parallel to the long axis of the body. There are also distinct, obvious differences in the pattern of wing veins that can be used to distinguish the house fly from the little house fly.

At a temperature of 75°F, eggs of the little house fly hatch in about 3 days and the larval period lasts about 11 days; pupation takes approximately 10 days.

After emerging as adults, female little house flies are reported to live an average of 24 days. Larvae (maggots) of little house flies are flattened with prominent lateral spines. They are light cream colored when first hatched and become darker brown as they mature. Like the common house fly, little house fly larvae feed on decaying animal and vegetable matter and excrement. Adult females usually do not enter buildings, but males often become abundant indoors and hover about aimlessly in the middle of rooms and in shaded outdoor areas. They can be found throughout a building rather than just in food serving and preparation areas. These flies can transmit similar disease and parasite organisms as the common house fly.

Management Guidelines for House Flies and Little House Flies. Successful control of the common house fly and the little house fly requires an integrated management approach. Insecticides, exclusion techniques, and traps are used to reduce the numbers of adult flies inside buildings. In addition, sanitation practices, exclusion techniques, insecticides, and natural enemies can be used outdoors to reduce egg production and control developing larvae.

Before attempting to control flies in buildings, gather some information about the infestation. Make sure the fly species is correctly identified. Find out how the flies are getting into the building and try to locate where they are breeding. Estimate the size of the population so that control measures can be evaluated; traps and sticky tapes are useful techniques for monitoring fly populations and can supplement visual observations. Observe the locations and density of fecal and regurgitated food spots in areas where flies rest. A simple way to monitor fly activity is to attach a small white card to a resting surface and observe the buildup of spots over a period of time (Figure 5-25). Use similar cards to follow up after control measures have been taken to evaluate control effectiveness.

Sanitation. Sanitation is the primary control method used to reduce house fly and little house fly populations. Suitable larval development sites must be eliminated by keeping interior and outside areas free of garbage, decaying

FIGURE 5-25.

A simple way of monitoring fly activity is to attach a small white card to the flies' resting surface. Activity is noted by the buildup of spots over a period of time.

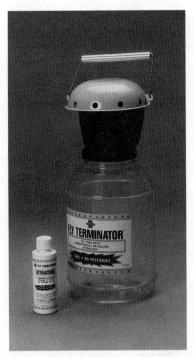

FIGURE 5-26.

Traps baited with an attractant can be used to capture large numbers of house flies in outside areas.

plant material, animal feces, and other food. Garbage cans are one of the main sources of domestic fly production in urban areas—as many as 20,000 fly larvae per week can be produced in a single garbage can in hot weather. Cans should be emptied twice each week and thoroughly cleaned of residues that support maggot growth. Use tight-fitting lids to keep adult flies from getting in and laying eggs.

Grass clippings allowed to decay in piles are another attractive source of food for fly larvae. Dispose of grass clippings, compost them, or spread them thinly over an area so they dry rapidly to avoid fly larvae buildup. Compost piles must be maintained and periodically turned and aerated so they will not become breeding sites.

Dog feces should be removed from yards daily and deeply buried or placed in sealed containers to reduce fly breeding sites. Livestock areas near buildings are also an important source of house flies. Animal manure should be removed from pens several times per week and composted to kill developing larvae.

Exclusion. Exclude flies from buildings by using screens over door and window openings. Air-curtains or fans above entrances to commercial buildings also help to prevent fly entry. Store garbage and other refuse in closed containers away from building entries to reduce the number of flies congregating there.

Trapping. Use traps baited with an attractant to capture large numbers of house flies (Figure 5-26). Commercial attractants are available. Fermenting molasses is a good bait for outdoor use; add ammonium carbonate to the bait to increase its attractiveness. Keep traps away from buildings. When traps become filled, bury or otherwise dispose of the dead flies.

Sticky paper traps (Figure 5-27), pheromone traps, and electrocutor traps work well indoors and do not emit obnoxious odors. Electrocutor traps use an ultraviolet light to attract flies to an electrically charged grid, killing them on contact (Figure 5-28). Electrocutor traps can effectively reduce adult fly

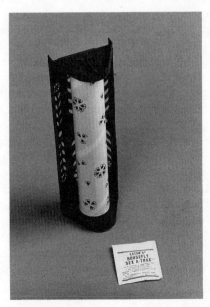
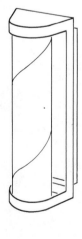

FIGURE 5-27.

Pheromone and sticky paper traps are useful for catching flies in confined areas. The packet contains pheromone that can be placed in the trap.

FIGURE 5-28.

Electrocutor traps use an ultraviolet light to attract flies to an electrically charged grid.

populations in enclosed areas. However, do not use these types of traps outside because they are not selective and may destroy many nontarget and beneficial insects. See page 000 for information on using light traps in buildings.

Parasites and Predators. Natural and biological controls are very important in fly management around livestock, but are less important in other areas. Several naturally occurring wasp parasites attack house fly pupae. Parasites can be purchased and released to augment naturally occurring species. Many species of insects, birds (including poultry), reptiles, and small mammals feed on fly larvae, pupae, and adults, and can be significant predators of these pests.

Chemical Controls. Always combine house fly insecticide use with other control methods. Because of their high reproductive rate, house fly populations can quickly develop insecticide resistance, so it is important to use any registered insecticide selectively.

The judicious use of poisoned bait in commercial or rural areas is effective in reducing numbers of adult flies. Apply bait around the outside of buildings, where garbage is stored near restaurants and other food preparation areas, and near livestock areas. Bait can also be placed in feeding stations; locate bait stations in areas of greatest fly concentration. Most fly baits contain a synthetic pheromone to attract adult house flies; others may contain sugar or molasses as an attractant. Avoid placing bait in areas where it could be a hazard to children or pets. Poisoned bait kills flies rapidly, but its effectiveness is short-lived and requires repeated applications.

If large numbers of adult flies appear in a building, first try to let them out through windows. If this is not successful, use a quick- acting, short-residual synergized pyrethrins insecticide for rapid knockdown (be sure the insecticide is labeled for the location where it will be used). Apply the spray as a mist in tightly closed rooms; never make applications to rooms while food is being prepared or to areas that cannot be thoroughly aired out before people or animals return. In food preparation areas, be sure preparation surfaces are covered and food and utensils are put away before making an application. Insecticide application only provides temporary control, however, so locate and destroy breeding areas and block the flies' entry into the building. As a last resort, treat breeding sites with an insecticide registered for control of larvae; however, this may enhance the development of insecticide resistance. Before using any treatment, determine if natural enemies are controlling part of the population; if natural enemies are present, they will be destroyed by the insecticide as well, resulting in a possible resurgence of the fly population to even higher levels unless larval food supplies are totally eliminated. When other methods fail to provide adequate reduction of fly populations, apply a residual insecticide spray to outdoor surfaces to control adult flies; spray overhead structures such as eaves, beams, wires, and ceilings of porches, patios, carports, garages, and breezeways, according to label instructions. Because house flies rest in these areas at night during warm weather, they will be killed by prolonged contact with the insecticide. During cooler weather, adult house flies and little house flies attempt to enter buildings to rest for the night, therefore spraying outdoor resting areas is less effective. Do not apply insecticides to outdoor furniture, play equipment, fences, walls, or other surfaces that children or adults might contact.

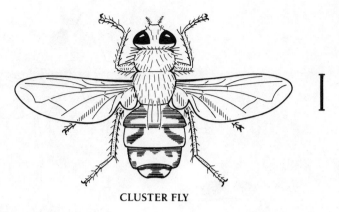

CLUSTER FLY

FIGURE 5-29.

Cluster fly, Pollenia rudis.

Cluster Fly
Pollenia rudis

Cluster flies (Figure 5-29) are nuisances in buildings where they congregate in the fall and through the winter. These flies do not carry disease organisms or eggs of parasitic worms, and are therefore not considered threats to public health.

Description, Development, and Habits. The cluster fly resembles the common house fly, but is larger and darker. The adult is about ⅜ inch long, with short, golden hairs on its thorax and a black and silver checkered pattern on its abdomen. When at rest, the wings of the cluster fly overlap, similar to the little house fly. Flight is sluggish and accompanied by an audible buzzing sound. These flies enter buildings by crawling through small openings in attics, siding, and under eaves. They get into rooms through gaps in moldings and baseboards, window pulley holes in older sash-type windows, and other small openings. During the fall, swarms of cluster flies accumulate in attics, closets, and empty rooms and leave stains on walls, draperies, and other surfaces where they rest. On warm, sunny days in early spring they are annoying because they fly around in rooms and collect on windows.

Cluster flies deposit eggs indiscriminately in the soil. After hatching, larvae search out earthworms, which they parasitize and feed on over a period of about 13 days. Larvae leave their hosts to pupate in the soil; pupation takes about 2 weeks. Four broods per year is typical.

Management Guidelines for Cluster Flies. The primary control of cluster flies should be to limit their access to buildings by closing off any small openings that provide entry. Use a labeled knockdown insecticide to eliminate flies in rooms when their presence cannot be tolerated; apply this as a mist in tightly closed rooms. Spray a residual insecticide, labeled for control of flies in buildings, onto surfaces of attics, basements, inside wall voids, and in confined areas where the flies are seen resting or are gaining entry into rooms. In late summer to early fall, use an appropriate residual insecticide around the outside of buildings to keep flies from entering. Direct the spray beneath eaves and roof gables and around windows.

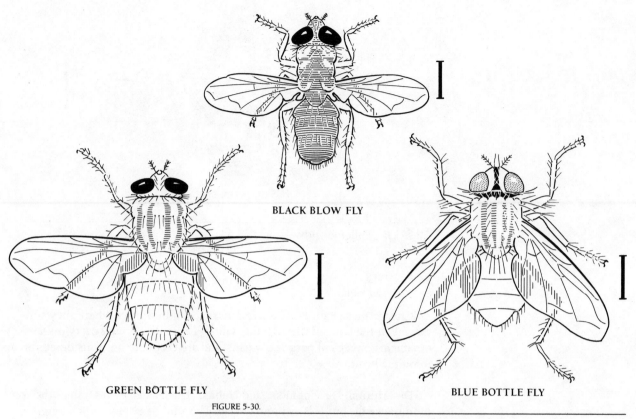

BLACK BLOW FLY

GREEN BOTTLE FLY

BLUE BOTTLE FLY

FIGURE 5-30.

Green bottle fly, Phaenicia (Lucilia) sericata *(left), black blow fly,* Phormia regina *(center), and blue bottle fly,* Calliphora vicina *(right).*

Black Blow Fly
Phormia regina

Green Bottle Fly
Phaenicia (Lucilia) sericata

Blue Bottle Fly
Calliphora vicina

The blow flies (Figure 5-30), including the black blow fly, the green bottle fly, and the blue bottle fly, are occasional pests in buildings. They are usually seen flying up against window glass or screens; these flies make audible buzzing sounds. Typically these flies emerge from rotting meat or carcasses of dead animals, so their sudden appearance in large numbers in a building could indicate the presence of a dead animal in a wall void, attic, crawl space, or other inaccessible area. Smaller numbers of blow flies in a building may indicate that a dead animal or rotting meat may be present in the neighborhood. Adult blow flies readily deposit eggs in exposed meat or fish in kitchens, markets, or other food-handling establishments and are capable of transmitting disease organisms to food in this manner. These flies are most active during warm, sunny periods and usually enter buildings in the spring and fall, seeking shelter from cool nighttime temperatures. Because they are strongly attracted to flesh, they attack an animal soon after death or begin depositing eggs into fresh meat a few minutes after exposure. Blow flies also deposit eggs into wounds of animals and people, resulting in a condition known as myiasis—the invading and consuming of living tissue by fly larvae.

Description, Development, and Habits. Blow flies are larger than the common house fly. Adults range between ⅜ inch and ½ inch in length. The black blow fly is a blue-green, dark blue, or greenish black color. The green bottle fly is brilliant metallic bluish green. The blue bottle fly has a metallic blue abdomen with a dark grayish thorax and large red eyes. These flies have four to eight broods per year and females produce up to 600 eggs, laid in batches of 100 to 200; eggs are laid in decomposing animal flesh or garbage containing some animal matter. Blow flies also deposit eggs into dog feces or any decaying organic matter with a high crude protein content such as dry cat food. The usual larval stage lasts 2 to 10 days; pupation takes place in the soil. Full-grown larvae can hibernate in the soil over the winter.

Management Guidelines for Blow Flies. The most important control for blow flies in buildings is exclusion. Use screens on windows and doors and seal up other openings where they might enter. Dispose of spoiled meat, animal byproducts, dog feces, and other waste products in and around buildings to avoid attracting blow flies. The sudden appearance of large numbers of blow flies inside a building usually indicates the presence of a dead animal, perhaps as a result of pest control activities directed toward rodents or other vertebrate pests. The dead animals should be removed if possible. Apply a labeled pyrethrin or pyrethroid spray as a fine mist in a tightly closed room for rapid knockdown of adult flies if population numbers are high.

Several traps are commercially available for trapping blow flies. These use an attractant and have an escapeproof entrance.

A simple way to reduce problems in noncommercial structures where blow flies are persistent pests and larval sources cannot be located is to make a small, pencil-sized opening at the top corner of each window screen so flies can exit; blow flies are attracted by the light coming in a window and habitually crawl to the top of window screens. Small holes such as these should not allow other pests to enter.

Vinegar Flies
Drosophila spp.

Vinegar flies (Figure 5-31) are also known as fruit flies or pomace flies. These pests are associated with overripe or rotting fruit and vegetables as well as garbage, waste water, and residues from beverages. Adults are small enough to gain entry into buildings through most window screens. They can also be carried into buildings as eggs or larvae on fruits and vegetables and other food items. Besides being a nuisance, vinegar flies can transmit diseases to people through food contamination. Vinegar flies also transmit bacterial and fungal organisms to noninfected fruit, which cause the produce to spoil. Adults of these tiny flies are often seen hovering over fruit or garbage. They are serious problems in packing houses and food processing plants as well as kitchens and restaurants.

Description, Development, and Habits. Adults are about ¹⁄₁₀ inch long and yellowish with dark crossbands on their abdomens. They have a distinctive feathery antennal bristle and some species have red eyes. Females produce about 500 eggs in a lifetime. During warm weather, their life cycle is about 8 to 10 days. Larvae feed on microorganisms, especially yeasts, growing on rotting or decaying food or on plant, fruit, or vegetable exudates. They leave the food source to pupate, preferring loose soil.

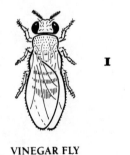

VINEGAR FLY

FIGURE 5-31.

Vinegar fly, Drosophila *spp.*

Management Guidelines for Vinegar Flies. Exclusion is of little value in controlling vinegar flies because of their small size. Sanitation practices are the most effective control methods used to reduce populations, so eliminate food and breeding sources including rotting or fermenting fruits and vegetables and residues such as peelings, cores, and juices. Thorough, daily cleanup must include the removal of any food items accidentally swept beneath appliances, tables, counters, or other equipment. Clean up spilled liquids thoroughly to prevent vinegar flies from breeding in dried residues. Remove liquids and food particles from cracks and crevices. Place garbage in plastic bags and keep it in covered trash cans well away from buildings. Dispose of dishwater, water from floor moppings, and garbage-laden water from sinks. Rinse mops in bleach and hang them to dry.

Use a knockdown spray registered for this purpose to reduce populations of flying adults in enclosed areas. Apply the spray as a fine mist, but avoid contact with food or food preparation and handling equipment; confine this spray to areas that cannot be cleaned thoroughly on a regular basis, such as under and behind appliances or fixtures. Synergized pyrethrins insecticides used as knockdown sprays have a short residual life, so these should be used along with a thorough sanitation program. Use traps baited with fruits and yeast to collect adults in enclosed areas, but combine trapping with careful sanitation methods. In areas around packing houses or food processing plants, where large volumes of culled produce or food wastes attract high populations of vinegar flies, spray insecticides directly on the waste to repel adults and retard egg deposition and larval development. Insecticides used this way provide about 24 hours of protection, allowing time for wastes to be collected and disposed of.

6 Parasitic, Biting, and Stinging Arthropods

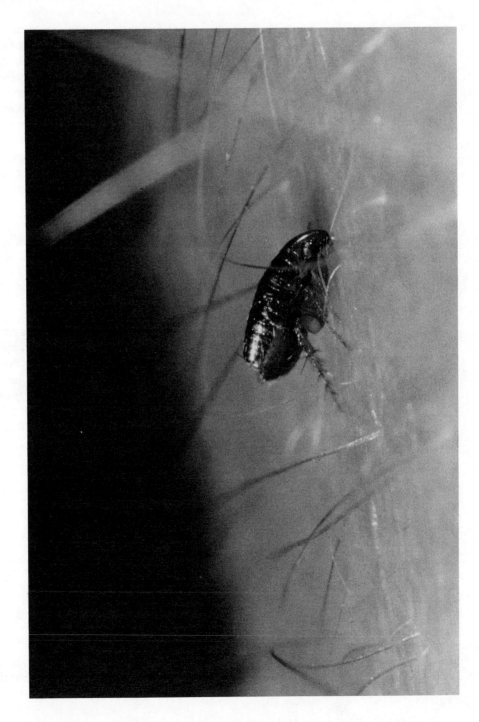

rthropods that bite or sting to defend themselves or those that feed on the blood of people or domestic animals are serious and sometimes dangerous pests. Bites or stings may result in localized painful itching and swelling which, if scratched, may even lead to a bacterial infection. Some arthropods transmit disease organisms through the wounds they cause. Some stinging and blood-feeding pests inject venoms capable of causing allergic reactions. In some people, these reactions can be fatal. Arthropods that feed on the blood of people or their pets include mosquitoes, some hemipterans (the true bugs), fleas, lice, ticks, midges, and some mites. Biting or stinging arthropods include bees, wasps, spiders, and scorpions in addition to ants, which are discussed in the previous chapter.

This chapter describes those parasitic, biting, or stinging pests that are most commonly found in or around homes or other buildings. The ones that may require pest control include mosquitoes, bugs, fleas, bees, wasps, mites, and spiders. Control of lice is usually supervised by medical or health professionals. Ticks in structures are usually associated with pets and are best controlled in cooperation with a veterinarian. Some ticks transmit the organism that causes Lyme disease, therefore people suffering tick bites should seek medical attention.

MOSQUITOES

Mosquitoes are blood-feeding pests of people and animals (Figure 6-1). These insects belong to the order Diptera and are related to house flies, gnats, and midges. Eggs of most mosquitoes are laid in water, usually on raftlike structures made by the adult females. Mosquito larvae are aquatic; they feed on algae, protozoans, and minute organic debris. Adult mosquitoes are winged and free living. Males do not feed as adults, but females of most species require a blood meal before oviposition, utilizing the protein in blood for egg

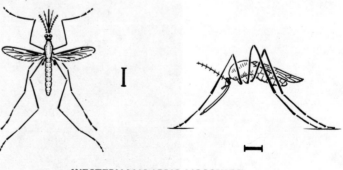

WESTERN MALARIAL MOSQUITO

FIGURE 6-1.

Western malarial mosquito, Anopheles freeborni.

Lyme Disease

Symptoms of Lyme disease were described in Europe over 100 years ago. The disease was named in 1975 by a physician studying the symptoms of a group of children living around Old Lyme, Connecticut. This disease appears to be spreading throughout most of the country and has been detected in at least 43 states. In California it seems most prevalent along the north coast, where over 300 cases were reported in Mendocino County alone during 1988.

The disease, when left untreated, can involve the brain, the joints, or the heart. It is caused by a corkscrew-shaped bacterium, or spirochete, similar to the one that causes syphilis. This spirochete is transmitted to people and animals through tick bites. In California, the western black-legged tick, *Ixodes pacificus*, is believed to be the only one of the 49 identified tick species in the state to transmit the spirochete. This tick is small and easily goes unnoticed. It occurs throughout most of California, usually in rural and forested areas.

Preventing tick bites is the best way keep from becoming infected with the spirochete. Anyone going into areas where the tick occurs should do the following:

- Wear light-colored clothing including long pants, a long-sleeved shirt, and a hat. Be sure as much skin area as possible is covered.

- To provide barriers to keep ticks from reaching the skin, pants should be tucked into boots or socks and shirts should be tucked into the pants.

- Spray a repellent containing DEET on any exposed skin. Spray clothing with the repellent or a product registered for use against ticks, such as permethrin.

- If possible, stay on clear paths and avoid trail edges, brush, and grassy areas.

- Examine all body areas for signs of ticks as soon as the clothing is removed. Shower immediately.

FIGURE 6-2.

Public health agencies usually have the responsibility of managing mosquitoes.

production. Mosquito bites usually result in red, swollen areas called wheals which itch severely and may persist for several days. Some people develop allergic reactions to proteins injected by mosquitoes and become ill after being bitten. Some species of mosquitoes also vector microorganisms to people or animals. These microorganisms include those that cause malaria, yellow fever, encephalitis, dengue, canine heartworm, and filariasis.

Management Guidelines for Mosquitoes. Mosquito control requires area-wide management of breeding sites. This type of management is usually the responsibility of public health agencies and mosquito abatement districts (Figure 6-2). Control of mosquitoes in and around buildings depends on sanitation to eliminate breeding sites and exclusion to keep mosquitoes out of buildings.

Sanitation involves draining any standing water and eliminating all objects or containers, such as old tires, cans, bottles, and dishes, that contain water where mosquito larvae can develop. Look for blocked rain gutters, for instance, as sources of standing water. Certain mosquitoes develop in water trapped in cavities of trees or basins formed by tree branches. Some species require only small quantities of water for short periods of time to develop.

Livestock and pet water containers should be emptied and cleaned regularly to prevent mosquitoes from using these as breeding sites. Fish ponds and other bodies of water that cannot be drained periodically should be stocked with the small mosquito fish (*Gambusia affinis*) that effectively controls developing larvae. Quantities of these fish can be obtained through local mosquito abatement districts.

Sanitation practices may need to be combined with chemical control for some situations. Treat water sources such as swimming pools, fountains, reflection pools, and decorative ponds with chlorine, copper sulfate, or other algaecides to eliminate conditions favorable to mosquito larvae development. Use a film of light vegetable oil in small puddles or ponds to cut off oxygen from mosquito larvae. A special strain of the bacterium *Bacillus thuringiensis* is effective as a mosquito larvicide. This material is applied to aquatic areas infested by larvae.

To control adult mosquitoes in localized outdoor areas, apply an appropriate insecticide in the form of a fog, using a fog applicator. Be sure the fog is dispersed throughout the treated area so it comes in contact with flying and resting adults. Apply fogs only when the wind is calm to allow the droplets to remain suspended in the atmosphere around the treatment site.

Exclude adult mosquitoes from buildings by using screens over doors, windows, and other openings. If necessary, use appropriate insecticides to reduce numbers of biting females inside enclosed areas. Reductions will only be temporary, however, unless steps are taken to exclude adult mosquitoes, eliminate breeding sites, and destroy larvae.

An insect repellent applied as a lotion or aerosol spray to skin or clothing reduces mosquito attacks on people who must spend time outdoors in areas where mosquitoes are a problem (Figure 6-3). Apply repellents frequently, according to the directions, for best results.

HEMIPTERANS

Hemipterans, or true bugs, comprise a large order of insects with a few species that occasionally attack people or are household pests. Insects in the order Hemiptera undergo incomplete metamorphosis, therefore wingless young resemble winged adults and usually have the same feeding habits. Some hemipterans, such as the bed bug and species of conenose bugs, require protein-rich blood from people or animals for development.

Bed Bug
Cimex lectularius

Bat Bug
Cimex pilosellus

Swallow Bug
Oeciacus vicarius

Bed bugs, bat bugs, and swallow bugs (Figure 6-4) are no longer common pests in most areas, but they may appear from time to time and require control. The bed bug is adapted to living in dwellings. A raised, wheallike bump appears on the skin where a bed bug feeds. There may be two or three closely spaced punctures at that site. Bed bugs also parasitize chickens, mice,

Insect Repellents

Insect repellents are special chemical compounds that repel biting insects, as well as ticks and chiggers. They do not appear to have an effect on stinging insects such as bees or wasps. To be sold in the United States, insect repellents must have EPA Registration and Establishment numbers.

There are over 15 major insect repellent products on the market. These are available in a variety of forms, the most common being the aerosol spray can. Repellents are also available in lotion form, in squeeze bottles, as rub-on sticks, and in foil-wrapped towelettes.

Two major active ingredients are used as repellents: ethyl hexanediol and N,N-diethyl-meta-toluamide (DEET). Manufacturers may use either of these ingredients or a combination of both. The effectiveness of the repellent is usually the result of the concentration of the active ingredient(s) in the ready-to-use formulation. Formulations containing higher percentages of active ingredient (up to 100%) are more effective for longer periods of time. DEET appears to be the most effective repellent of common biting insects and ticks, and is the recommended ingredient to protect against the western black-legged tick that vectors the spirochete causing Lyme disease.

Some repellents remain effective for less than an hour, and others may last for 10 hours. Factors contributing to the lasting effect of repellents include sweating, swimming, rain, and general level of activity, the concentration of active ingredient, and other ingredients in the formulation.

FIGURE 6-3.

An insect repellent applied to the skin or clothing usually reduces mosquito attacks on people who must spend time outdoors.

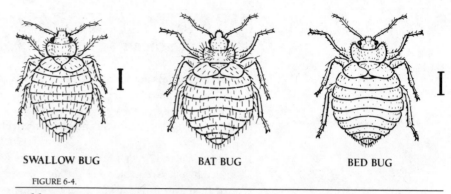

SWALLOW BUG　　　BAT BUG　　　BED BUG

FIGURE 6-4.

Bed bug, Cimex lectularius *(right), bat bug,* Cimex pilosellus *(center), and swallow bug,* Oeciacus vicarius *(left).*

rats, and rabbits. The bat bug and swallow bug occasionally enter dwellings and, given the opportunity, feed on the blood of people.

Description, Development, and Habits. The adult bed bug is less than ¼ inch long, oval, flat, and rusty red or mahogany. Smaller, immature bed bugs are yellowish white. Adults are flightless, having vestigial forewings that extend barely to the base of the abdomen. After a blood meal, the abdomen becomes rounded and distended. Females lay between 200 and 500 eggs, in batches of 10 to 50, in cracks or crevices of protected places. Eggs are cemented to the surface where they are laid; nymphs emerge from eggs in about one week. Nymphs pass through five instars, or growth stages, before becoming adults; they require a blood meal to pass into each successive instar. Females also take a blood meal before laying each batch of eggs. It takes from 3 to 10 minutes for a bed bug to complete a single meal. Between feedings, bed bugs return to their hiding places in cracks or crevices to molt or lay eggs. They travel considerable distances from these nesting sites to locate suitable hosts. Bed bugs are nocturnal, so they search for hosts and take blood meals while people are sleeping.

The average development time from egg to adult ranges between 37 and 128 days, depending on temperature. Under normal feeding conditions, female bed bugs live for about 315 days. If suitable hosts are not available, the bugs are able to survive for long periods without food, which, according to some studies, can increase their life span to as much as 550 days.

The bat bug and swallow bug are slightly smaller than bed bugs, but similar in appearance. They may occasionally feed on blood of people when their normal hosts (bats or swallows) nest near or within a building and nesting sites have recently been disturbed or removed.

Management Guidelines for Biting Bugs. Bed bugs are reclusive and hide most of the time in small cracks and crevices. Look for resting bugs in upholstery, behind loose wallpaper, in picture frames, around moldings, in and among furniture, and any other place that provides a suitable habitat. Check for fecal spots (brown, yellow, and black spots on walls and linens), which might be apparent even though bugs cannot be seen. Other signs to look for include empty egg cases and exuviae, or molted "skins."

Begin control with thorough sanitation practices to remove debris and other objects that serve as bug nesting sites. Repair cracks and caulk gaps and openings in walls and moldings to eliminate as many hiding places as possible. Before caulking, treat these areas with a residual insecticide labeled for this purpose; this kills any bugs present and destroys nymphs that may hatch from eggs. Repair loose wallpaper where bed bugs can hide (Figure 6-5). Disassemble beds and thoroughly launder bedding and mattress pads and inspect mattresses and box springs for holes or tears that could harbor bed bugs; destroy damaged mattresses or box springs. If necessary, spray bed frames, mattresses, and box springs with a registered insecticide containing pyrethrins or pyrethroid—do not use residual materials and do not get fabrics wet. Apply the insecticide early in the day to allow ample time for the mattresses and box springs to air before being used. Use diatomaceous earth or silica aerogel dusts in inaccessible hiding places that cannot be caulked or sealed. Make additional insecticide applications periodically to destroy newly hatched nymphs.

Concentrate bat bug control in attics and other areas in the building where bats are located. When working in these areas, wear protective clothing and respiratory devices to avoid contact with bat feces or urine, which may contain disease organisms. Do not breathe dust from bat guano. Do not apply insecticides on or near bats; injuring or killing bats is illegal. Confine residual insecticides only to areas where bugs travel to rooms inhabited by people, such as through light fixtures, wiring or plumbing channels, tops of wall voids, and gaps or cracks in the construction. Another control option is to spot-treat the roosting area with a total-release aerosol, but only if bats are not present. Aerosols may kill roosting bats, which is prohibited by law. See the section on bats in Chapter 10 for information on bat control and precautions when working around bats.

Swallow bugs usually remain in the swallow's mud nests located under eaves and rafters of buildings. They leave nests and search for other hosts when the swallow population migrates or is destroyed. They also leave when their own population becomes so large there are insufficient numbers of natural hosts. Control involves removal of nests and treatment of nesting locations with a suitable residual insecticide. Be sure the insecticide is sprayed into cracks and crevices where swallow bugs enter buildings. An alternate method is to seal openings with caulking.

Because swallows are protected by federal and state regulations, contact the nearest office of the Department of Fish and Game before removing nests or applying any type of pesticide in the vicinity of occupied nests. See Chapter 10 for information on preventing swallows from nesting on buildings.

Western Bloodsucking Conenose
Triatoma protracta

The western bloodsucking conenose is an occasional pest in homes and other buildings (Figure 6-6); however, it is more common in rural areas. It occasionally feeds on people, causing severe local skin reactions as well as possible allergy problems. This bug belongs to the assassin bug family Reduviidae, sometimes referred to as kissing bugs. Conenose bugs feed on blood, like bed bugs, and are vectors of the trypanosome parasite that causes Chagas' disease in people, a serious problem in Central and South America. Some conenose bugs in the United States have been found to be infected with this trypanosome organism, but the incidence of Chagas' disease in the United States is low, probably because the bugs in this country are a different species.

Description, Development, and Habits. The western bloodsucking conenose is between ½ and ¾ inch long and dark brown to black. It occasionally has a narrow tan edge along its lateral abdominal margin. This bug passes

FIGURE 6-5.

Loose wallpaper can provide hiding places for bed bugs.

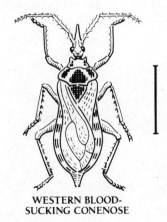

WESTERN BLOOD-
SUCKING CONENOSE

FIGURE 6-6.

Western bloodsucking conenose,
Triatoma protracta.

through five nymphal instars, all of which are usually active throughout the year. Adults live as long as 3 years and are the life stage usually found in buildings. The western bloodsucking conenose normally lives in nests of wood rats, pack rats, or other rodents and feeds on rodent blood. Conenose bugs may be attracted to outdoor lights on warm nights, promoting their entry into buildings. The occasional attack on people usually takes place during the night while a person sleeps; rarely is the bug discovered while it feeds. Several punctures are made about ¼ inch apart along a straight line at the feeding site. This injury may swell and become painful for several days to a week or more. In buildings, this insect hides in cracks and concealed areas during the day.

Management Guidelines for Conenose Bugs. Control of the western bloodsucking conenose involves sanitation practices around buildings to remove rodent nests. Exclude bugs from buildings with screens and by caulking cracks and other openings. To keep from attracting these insects at night, do not use outdoor lights around doorways. Examine your pets carefully before letting them in from the outside, especially at night, because they may carry bugs into the building. If western bloodsucking conenoses are established in a building, follow the control guidelines described above for bed bugs.

When conenose bugs invade a building and threaten injury to people, it may be necessary to eliminate them with an insecticide. Use a registered total-release pyrethrins space spray or a registered quick-acting, short-residual fumigant. Be sure to treat all rooms where these bugs may be hiding.

FLEAS

Fleas belong to the insect order Siphonaptera. They are tiny wingless insects that undergo complete metamorphosis, having egg, larval, pupal, and adult stages (Figure 6-7). Adult fleas are parasites of warm-blooded animals. There are over 1600 described flea species in the world, 95% of these associated with mammals, although people are the only primate host; the remaining species parasitize birds. Most flea species have a host preference, but some attack other hosts for blood if necessary.

Some flea species are capable of transmitting serious disease organisms, such as bubonic plague or murine typhus, either through their bite or

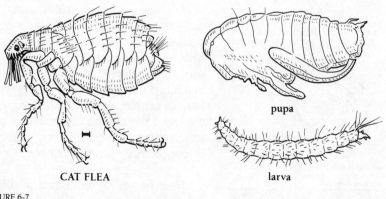

CAT FLEA

pupa

larva

FIGURE 6-7.

Fleas undergo complete metamorphosis through larval, pupal, and adult stages.

through their feces. The Oriental rat flea is the primary vector for the plague-causing bacterium, *Yersinia pestis.* This organism is found in rodent populations throughout temperate areas of the world, including California. Some other flea species can also transmit this bacterium. Murine typhus is caused by the microorganism *Rickettsia mooseri,* which occurs in rodent populations and can be vectored to people by fleas. Certain fleas are also intermediate hosts for dog and rodent tapeworms, intestinal parasites which can be transmitted to pets and people through ingestion of infected fleas.

Through their feeding, fleas inject a hemorrhagic saliva which destroys tissues and causes bleeding at the feeding site. The bite produces severe itching, and multiple bites may cause a generalized rash. Some animals and people are more sensitive than others to flea bites and may experience allergic reactions. However, the reactions may be delayed for periods of time, making it difficult to associate the discomfort with the cause. People or animals who suffer from flea allergies are very uncomfortable and develop itching and skin rashes, often from just a single flea bite.

Several species of fleas may be encountered in the course of pest control in and around buildings. These include the human flea, *Pulex irritans,* the oriental rat flea, *Xenopsylla cheopis,* and the northern rat flea, *Nosopsyllus fasciatus.* The two most common fleas found on pets, however, include the cat flea, *Ctenocephalides felis,* and the sticktight flea, *Echidnophaga gallinacea.* Cat fleas are the most troublesome flea pests in buildings and dwellings. This will be the only species to be described here.

Cat Flea
Ctenocephalides felis

The cat flea is the most common flea pest worldwide. This species is a parasite of cats and dogs as well as opossums, foxes, coyotes, and other wild animals. It also attacks people. The cat flea is most active during warm summer weather, although it can be a pest throughout the year. In California, it is most abundant from May through October. Cat fleas have been identified as an occasional vector for the plague bacterium, *Yersinia pestis,* and they can also vector the organism causing murine typhus. Cat fleas are an intermediate host for tapeworms; treatment for tapeworm, therefore, requires that cat fleas be controlled.

FIGURE 6-8.

Only adult forms of fleas feed on liquid blood. A blood meal from a warm-blooded animal is necessary for females to produce eggs.

FIGURE 6-9.

Larvae of cat fleas are small and hairy with a whitish appearance.

Description, Development, and Habits. Adult cat fleas are about 0.08 inch long and dark reddish brown to black. Their laterally compressed bodies are an adaptation that enables them to pass freely through dense hair of host animals; their smooth cuticle is equipped with short, spinelike hairs directed backward which help them move forward rapidly through a host's coat. Cat fleas have long powerful legs and are capable of jumping considerable distances for their small size. They have been observed to jump 8 inches vertically and 15 inches horizontally.

Only adult fleas feed on liquid blood (Figure 6-8). A blood meal from a warm-blooded animal is necessary for females to produce eggs. However, unlike mosquitoes, both male and female fleas are parasites. Over their lifetime, female cat fleas produce up to 450 eggs, which are laid in batches of 3 to 18 at a time. Eggs may be deposited among the hair of the host or off the host in debris. Eggs deposited on a host usually drop off before hatching. Depending on environmental conditions, eggs hatch in 1 to 6 days; optimum conditions include warm temperatures and high humidity. Eggs hatch into small, hairy, whitish larvae having a distinct brown, eyeless head (see Figure 6-9).

FIGURE 6-10.

When immature fleas complete the larval stage, they pupate inside a silken cocoon. The cocoon becomes covered with debris, making it difficult to distinguish from other debris.

These pass through three instars and grow to about 3/16 inch long. Cat flea larvae feed on digested blood in the dried excrement of adult fleas to complete development. Places where the host animal sleeps usually have a high concentration of flea droppings. A protected, moist environment is required for larval development; relative humidity conditions below 75% are often fatal.

Carpets provide ideal conditions for cat flea larvae. Carpets installed directly over concrete floors are especially suitable because the cooler temperature of the concrete increases humidity. The sleeping pet provides warmth and also adds to the humidity.

Larval stages under optimum conditions last between 7 and 21 days, but during cool weather this period may be longer. Upon completion of the larval stage, immature fleas spin a silken cocoon which becomes covered with debris, camouflaging them from predators (Figure 6-10); they pupate into adults inside this cocoon. The normal pupal period lasts 7 to 14 days, although under extreme environmental conditions it may take longer. Once they become adults, fleas may remain inactive in the pupal case for up to a year, protected from enemies and adverse environmental conditions. Increased temperatures and direct physical contact stimulate adult fleas to emerge from the cocoon. Emergence may only take a matter of seconds. Emerged adults generally live for 25 to 60 days.

Management Guidelines for Cat Fleas. Controlling cat fleas in buildings is difficult and requires a multifaceted approach. Control methods must be coordinated with thorough, systematic cleaning of the infested area as well as efforts directed at eliminating fleas on pets.

Before starting a control program, survey the building to determine the sources of infestation (such as pets) and areas where breeding is occurring (such as pet sleeping areas). A sudden onset of flea complaints may indicate that a recent change has occurred affecting the flea population; for example, a cat died or is being boarded away from the area. Or the building may have just become occupied after being vacant. Once they become adults, fleas are mobile and search out new hosts when animal sleeping or resting sites are vacated, destroyed, or become overpopulated with other fleas. For instance, dogs and cats may crawl under buildings during warm weather to rest in the cool dirt; this may be an ideal area for the buildup of a large flea population.

In an infested area, flea populations are highest on rugs and furniture where dogs or cats regularly sleep; however, flea larvae are not usually found

TABLE 6-1

Ways to Reduce Fleas on Pets that Live in Buildings.

TECHNIQUE	EFFECT
PET CARE	
Bathe pet on a regular basis (two to four times a month).	Kills some fleas. Dislodges loose hairs and skin debris that serves as food for larvae.
Groom pet daily, using fine-toothed flea comb. Good technique for cats.	Removes adult fleas and eggs. Removes loose hairs and skin debris.
Confine pet to single indoor sleeping area.	Keeps fleas confined to localized area where control efforts can be concentrated.
Spray pet and sleeping area with flea repellent. (Some repellents last 30 to 60 days, others must be applied as frequently as once each day.)	Helps to reduce number of fleas attacking pet.
Consult veterinarian for flea control product for use on pet. Many products cannot be used on cats.	Flea control products repel or kill fleas coming in contact with pet.
INTERIOR HOUSEKEEPING	
Vacuum areas where pets sleep or spend time on a regular basis. (Clean at least twice per week and immediately dispose of vacuum cleaner bag.)	Removes some eggs, larvae, adult fleas, and skin debris. Also removes adult flea excrement and dried blood that provides food for larvae.
Keep pets out of carpeted areas and other hard-to-clean areas (such as closets).	Makes housekeeping functions that reduce fleas easier to perform.
Launder pet bedding on a weekly basis.	Kills eggs and larvae. Eliminates skin debris and hair.
EXTERIOR MAINTENANCE	
Mow grass, destroy weeds, and trim shrubbery. Perform weekly as needed.	Exposes eggs and larvae to more sunlight and kills them.
Irrigate areas surrounding buildings on a regular basis.	Kills eggs and larvae.

in areas of heavy pedestrian traffic, locations that receive exposure to sunlight, or areas where adult flea feces containing dried blood is not present. Flea eradication in a building cannot be permanent unless fleas are controlled on animals that spend time in the building (see Table 6-1).

Areas where adult fleas, flea larvae, and flea eggs are found must be cleaned thoroughly and periodically. Occupants should vacuum floors, rugs, carpets, upholstered furniture, and crevices around baseboards and cabinets at least weekly to remove flea eggs and larvae and food sources. Vacuuming is very effective in picking up adults and stimulating preemerged adults to leave their cocoons. Vacuuming must be repeated frequently because disturbed flea larvae quickly attach to carpet fibers, withstanding the pull of the vacuum. Fleas can survive and develop inside vacuum bags and adults may be able to escape into the room, so destroy bags by burning or by sealing them in a plastic trash bag and placing them in a covered trash container. Pet bedding should be laundered in hot, soapy water at least once each week.

Several insecticides are registered for controlling fleas, although some are limited to outdoor use only. Fleas are known to build up resistance to insecticides, so plan their use carefully and include pesticide use with other methods of control such as thorough, frequent vacuuming. Application techniques are important to ensure adequate coverage. Use the correct volume of spray per unit of treated area so that a lethal dose reaches the adults and larvae; follow label instructions carefully. If immediate control of adult fleas is needed, use a short-residual knockdown spray labeled for this purpose. Spray carpets, pet sleeping areas, baseboards, window sills, and other areas harboring adults or larvae. Combine the knockdown spray with an insect growth regulator (IGR) to prevent immature stages from developing into adults; this provides long-term flea control. The IGR should be applied as a diluted space spray or as a residual fog. In some cases the IGR can be applied without the knockdown spray, especially early in the spring or summer. However, elimination of adults may take several weeks. Vacuuming should be used with the IGR treatment program to destroy adults.

Prevent pets from resting in areas under buildings to eliminate these locations as sources of infestation. Once pets are excluded, treat the soil with a persistent insecticide to kill adults, larvae, and larvae hatching later from eggs.

Dogs or cats living in a building should be treated for fleas at the same time the building is being treated. This will slow down the rate of reinfestation. Pet owners should contact their veterinarian for advice and assistance in controlling fleas. It is important to know the names of insecticides being applied in or around the building where pets live. If carbamate or organophosphate insecticides are part of a building flea control program, pet owners should not use similar materials on their animals. Pets treated with organophosphates or carbamates will be more sensitive to these insecticides in their living areas. Pet products that contain organophosphate or carbamate insecticides include some brands of flea spray, flea powder, flea collars, materials used to control ticks, and internal medications given to animals for control of fleas or internal parasites or for protection against dog heartworm. Find out if any of these materials are being used before making any insecticide treatment.

Pet Flea Control Products

Several types of products are available to control fleas on dogs and cats. To protect the health of the animal and to ensure adequate control, it is advisable to consult a veterinarian before using any of these products. The effectiveness of most flea-control products usually depends on the thoroughness of application, frequency of application, amount of active ingredient being applied, the concentration of fleas in the animal's living quarters, and sources of reinfestation. Successful flea control on an animal must always involve controlling all life stages of fleas living off the animal as well. Young children and infants should not be allowed to contact pets who have been sprayed, dusted, or dipped with flea control products or who are wearing flea collars.

Soaps and Shampoos: Soaps and shampoos for flea control contain small amounts of insecticide (usually pyrethrins) that kill fleas on the animal. These require that the animal's body be thoroughly coated with a lather, and often it is recommended that the lather be allowed to remain on the animal for up to 15 minutes before rinsing. These products are most suitable for dogs. The bathing process assists in removing dried blood and skin flakes that provide food for flea larvae in the animal's sleeping area. This type of treatment does not usually have any lasting effect on keeping fleas off the animal. Wear rubber gloves to avoid contacting the insecticide while bathing the animal.

Repellents: Repellents are materials that are sprayed or wiped onto the animal's coat to repel fleas. Repellents may be combined with flea sprays or other flea control products. These last for only a short period before they lose their effectiveness through the animal's grooming and other activities and natural breakdown of the chemical repellent. They are most useful to protect an animal that will be going into a flea-infested area for a short time.

Dips: Dips are insecticidal liquids that must be diluted with water. The animal is submerged into the water to provide uniform application. Properly applied dips provide immediate control of fleas on the animal and will kill fleas coming in contact with the animal's coat for several days or longer. Avoid contact with the animal's eyes, nose, mouth, or ear canals. Wear waterproof gloves and avoid getting the liquid onto your clothing when handling the animal. Dips can be very hazardous to animals being treated if the insecticide is not properly diluted. This type of treatment should be performed by or under the supervision of a veterinarian.

Powders and Dusts: Powders and dusts are ready-to-use formulations consisting of a small percentage of insecticide active ingredient combined with inert powder. They also include desiccants such as silica aerogel which may or may not also be combined with a small amount of insecticide active ingredient. Sprinkle the pet with the dust and work it into the coat by brushing. Most powders are available in a shaker can, and some are available as aerosols. Wear rubber gloves when applying the dust. Avoid breathing any dust. Be extremely careful to avoid getting dust into the animal's eyes or nose. Powders and dusts are suitable for flea control on dogs and cats. They usually provide up to one week of control depending on the pet's activities and whether the animal gets wet.

Spray-On Liquids: Spray-on liquids, among the most common type of flea control product for pets, are ready-to-use formulations packaged in aerosol cans, squeeze bottles, or pump-type applicators. They contain a small percentage of one or more insecticides dissolved in a petroleum solvent. The animal's coat must be thoroughly wetted with the spray for it to be effective. For animals with very dense fur, it is necessary to brush the coat up while applying the spray. Wear rubber gloves during application. Avoid contact with the liquid spray. Do not get any spray into the animal's eyes, nose, mouth, or ear canals. Some spray-on products provide flea control for 30 days or more under certain conditions and when properly applied. These impregnate the animal's hair and slowly release the active ingredient.

Flea Collars: The plastic band on a flea collar is impregnated with an insecticidal material that is released slowly as a vapor while the collar is being worn. A collar may be effective for several months as long as it remains dry. Flea collars are suitable for use on dogs and cats; however, some animals are sensitive to the insecticide and may develop a rash or sores on the skin beneath the collar. They should not be used on animals showing any sensitivity.

Systemics. Systemic flea control products are administered on a regular basis in the form of a pill as an internal medication. They contain insecticidal materials that are transported to all skin areas through the animal's blood. The dosage administered to the animal is very critical and is based on the animal's body weight. Systemics should only be prescribed by a veterinarian, and the animal must be regularly monitored by the veterinarian for any adverse effects.

Once an area has been treated for cat fleas, periodically monitor for reinfestation. Remember to thoroughly vacuum floors, carpets, and furnishings on a regular basis. Carpeted areas where pets sleep should be vacuumed daily or at least every other day. Pets must be inspected and treated to remove fleas regularly. Pet bedding should be laundered weekly—more often in warm weather. Thoroughly clean items brought into the building, such as used carpets or upholstered furniture, to prevent these from being a source of flea infestation.

BEES AND WASPS

Bees and wasps, like ants, belong to the insect order Hymenoptera. They pass through a complete metamorphosis. Adult bees and wasps are nectar feeders, although some adult wasps paralyze insects or spiders as food for their larvae and may consume small amounts of their prey's blood before stocking their nest. These insects are generally considered to be highly beneficial, although bees and wasps can be nuisances around buildings because they forage for food among flowers and around outdoor dining areas. They occasionally wander indoors. Sometimes bees or wasps become nuisances when they construct nests in wall voids, attics, and other areas in or near buildings. Their nests can also be the source of carpet beetle infestations.

Bees and wasps are most notorious—and feared—because they defend themselves with a painful, venomous sting. The sting usually produces an intense local reaction accompanied by varying amounts of swelling. In some sensitive people, bee or wasp venoms evoke severe allergic reactions, known as anaphylaxis. After being stung a few times, allergic individuals may become hypersensitive to the venom's complex amino acids, proteins, and enzymes. Occasionally the reaction is so severe that a sensitized person may die shortly after receiving a sting unless drugs to counteract the allergic effects are administered.

Considerable interest and concern is being expressed regarding the migration of the Africanized bee into the United States. This is a much more aggressive strain of honey bee that is more prone to attack people. It has been responsible for several human deaths. Victims of Africanized bee attacks have received extremely large doses of venom as a result of hundreds of stings.

Description, Development, and Habits. Honey bees, the social wasps, and a few solitary wasps are the pests of this insect group most likely to appear in and around buildings. The social wasps and honey bees are more serious problems because their nests contain hundreds of individuals that may readily become aggressive when disturbed.

Honey Bee
Apis mellifera

Adult honey bees (Figure 6-11) range in length between ½ and ¾ inch and have two pairs of wings. They are black, gray, or brown, intermixed with yellow, sometimes with yellow-banded abdomens. Honey bees have a covering of fine, short hairs on the thorax, legs, and abdomen. Several different subspecies, or races, of honey bees exist, each having slightly different colors and other distinguishing physical and behavioral characteristics. Honey bees are social insects and their colonies are divided into three castes of individuals: the queen, the workers, and the drones. Queens are the reproductives for a colony and are larger than workers (Figure 6-12); unlike many of the ant

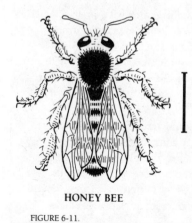

HONEY BEE

FIGURE 6-11.

Honey bee, Apis mellifera.

The Africanized Honey Bee*

The Africanized honey bee is slowly migrating northward out of South America, through Central America and Mexico. The migration of this highly aggressive strain of honey bees is expected to reach California during the 1990s, although researchers predict that it will take at least an additional five years after arrival in California before they begin to overwhelm European honey bee populations.

The main difference between Africanized honey bees and European honey bees is how they behave at the hive. Colony defense is the greatest concern. Tests repeatedly have demonstrated that Africanized honey bees alert to disturbances more quickly, prepare for colony defense more quickly, usually sting at least ten times more than European honey bees, and tend to continue the attack for longer periods of time and at much greater distances from the hive.

Experience suggests that elderly people with high blood pressure and weak hearts are at highest risk of death from Africanized honey bee stings. Small children are also more susceptible to the venom. Most older children and adults survive multiple sting incidents with limited physiological harm.

Serious injury or death can be prevented by prior planning if a person works in or around fields or other areas known to contain colonies of Africanized honey bees. Observe the following precautions:

1. Have a bee veil with you. Numerous stings can be tolerated elsewhere on the body, but your face should be protected.

2. Determine in advance where to seek shelter if bees begin to sting. A vehicle or building offers the best protection. Trying to lose the bees by hiding in foliage or walking among trees will not work.

3. If someone is stung and develops an allergic reaction to the venom, including becoming unable to breathe, a shot of the drug epinephrine is all that can save them. This medication must be administered immediately. Before entering areas where there is a likelihood of encountering Africanized honey bees, consult with medical authorities for information on obtaining and using epinephrine and for precautions and hazards associated with its use.

*Adapted from a paper given by Eric C. Mussen, Extension Apiculturist, University of California, Davis, on January 25, 1990, at the UCIPM Statewide Pest Management Seminar, Holtville, California.

FIGURE 6-12.

Honey bees are social insects that live in colonies, or hives, which may contain thousands of individuals. Queen bees are larger than the workers.

species, honey bee colonies usually have a single dominant queen. Nearly all of the bees in a colony are workers, nonreproductive females that defend a colony, perform housekeeping chores, forage for food and water, and feed and groom larvae, the queen, and drones. Drones, as the male bees are called, are slightly larger than workers and make more noise when they fly; they do not have stings.

The queen lays eggs in individual wax cells constructed by workers. Upon hatching, larvae are fed and groomed by workers throughout the larval stage. Food consists of honey, produced by workers from nectar collected during foraging flights to flowers, and pollen. When a larva completes its growth, workers cap its cell to prepare it for pupation. Upon completing pupation, the newly emerged adult chews its way out of the cell. Cells are cleaned and reused for subsequent brood.

Honey bee colonies may contain 7,000 to 60,000 individuals. Despite their venom and occasional pest status in or around buildings, honey bees are extremely beneficial because they serve as important pollinators to many agricultural crops. They are kept in hives by beekeepers and rented to growers for crop pollination. Honey and beeswax are extracted from hives, both of which are useful commodities.

Social Wasps

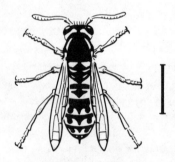

YELLOW JACKET

FIGURE 6-13.

Yellow jackets (family Vespidae) are social wasps. Also included in this group are hornets and paper wasps.

Many different species of yellowjackets, hornets, and paper wasps are included in the family Vespidae; species range in length from ½ to ¾ inch (Figure 6-13).

Paper and Umbrella Wasps

Paper wasps or umbrella wasps (belonging to the genus *Polistes*) are black with various yellow, orange, or greenish markings, depending on the species. Mated females overwinter in protected cracks or crevices, often in buildings. In spring they begin constructing paper nests from wood, which they chew into pulp (Figure 6-14); nests are constructed in attics or under eaves, in wall voids, in shrubs and trees, in cavities in the ground, and in lumber piles. Eggs are laid in a few cells and tended by the solitary female. When these offspring become adults, the original female assumes the role of queen and continues laying eggs while younger females forage for food and enlarge the nest.

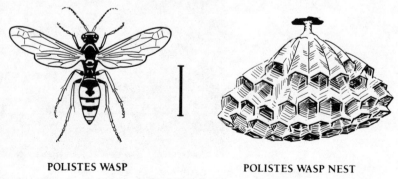

POLISTES WASP   **POLISTES WASP NEST**

FIGURE 6-14.

Paper wasps construct paper nests from wood which they chew into pulp.

Yellowjackets and Hornets

Adult yellowjackets and hornets (*genus* Vespula) forage for food, which they consume themselves or carry back to the nest for developing young. These are beneficial insects that contribute to the natural control of many plant pests. They become pests when their foraging brings them too close to people. In addition to capturing insects, they feed on fruit, juice and soft drinks, dead animals, feces, and meats. Some species are attracted to homes, buildings, and outdoor eating areas for foraging and nest building. Yellowjackets generally construct their nests in holes in the ground or in hollow logs or tree stumps near the ground. Hornets build large globular paper nests suspended from limbs of trees or shrubs or building overhangs. Most colonies die out toward the end of the first year, although some colonies may last for 2 years. Newly emerged females mate and overwinter to begin fresh colonies the following year.

Management Guidelines for Bees and Social Wasps. Destroying or removing bee and wasp nests is a delicate operation. It requires special equipment, protective clothing, and skill to prevent stings. Swarms of bees, which are most common in the spring, may appear on or near buildings and create a hazard to people. Because they are beneficial, bees should be removed rather than destroyed, if they are accessible. Whenever possible, obtain the assistance of a beekeeper for removing a hive in or near a building because beekeepers have the skills and equipment necessary to do a safe and thorough job. Remove the colony as soon as possible. Colonies that are located in wall voids or other inaccessible places are more difficult to deal with. The following box describes one method that can be used if time allows. To destroy a bee colony in a wall void or inaccessible location, use a fast-acting insecticide labeled for this purpose. Be sure all openings to the outside or interior of the building are sealed except the one used to apply the insecticide. Once the colony is killed, it must be removed from the building to prevent odors from dead bees and fermenting honey from permeating the building and to keep from attracting other insects such as ants to the nest. Old nesting sites are extremely attractive to new bee swarms, so all potential entrances ¼ inch or larger in the area of a building where a nest was removed must be sealed. Naphthalene (moth balls) may be useful in repelling bees from areas where nests once existed.

Method of Removing Bees Nesting in the Wall of a Building

Continuous honey bee flight activity to and from a hole in a building suggests there is a colony of bees inside. Often, the bees can be heard buzzing in some interior location.

Removing colonies of honey bees from buildings often is a complex task which is best undertaken by professionals. An experienced beekeeper usually can remove bees and combs from easily accessible places, but often bees live between walls or tucked away where they are impossible to reach.

Simply killing the bees *in situ* with an insecticide can have serious consequences:

Dead bees may form a deep, moist pile which promotes decay and produces strong odors. Dead bee brood produces equally offensive odors.

In the absence of bees, stored honey may absorb moisture and ferment or overheat. This results in burst cappings, and the freed honey may penetrate ceilings or walls, causing stains or sticky puddles around doors and windows.

The quickest way to remove bees is to kill them and then clean out the whole hive area. If an inner wall, ceiling, or outer wall must be removed, the services of a building contractor may be required. It is essential to remove all honeycomb and to plug all holes to be certain there is no access to the area. Remaining bits of beeswax emit an odor which is highly attractive to swarming bees.

A much slower method of removal may be used if there is no urgency involved and if it is desired to protect the bees and the building. This method is based on the principle of a one-way exit—bees which leave the building cannot return inside. However, the bees will cluster in a large mass around the exit, so an elaborate set-up is used to transfer the clustering bees to a new hive. Experienced beekeepers can do the job best, as they are used to bees flying around and to being stung occasionally. If you do it yourself, follow these steps:

- From a beekeeper, obtain a one-story hive containing one frame of unsealed brood covered with bees, one frame of honey, and adequate frames of drawn comb or foundation to fill the hive.

1. Fold a piece of window screen to make a cone wide enough at the bottom to cover completely the bees' entrance to the building. The cone's smaller opening should be about ⅜ inch in diameter. Bend the cone slightly upward toward the smaller opening.

2. Plug all other holes where bees may enter the building.

3. Protecting yourself with a bee veil and long sleeves (bee gloves are optional), use a bee smoker to smoke the bees' entrance to the building. Then fasten the large end of the screen cone tightly over the entrance.

4. Position the one-story hive near the cone entrance, possibly on brackets nailed to the building. Place the frames with brood and honey in the center of the hive; place frames of drawn comb or foundation at the sides. The hive entrance should be reduced to about a 1-inch opening to protect the colony from being robbed by stronger colonies that may be in the area. Bees emerging from the screen cone will not be able to find their way back through the cone opening. Instead, they will enter the new hive near their old entrance and become established there. As bees leave the building and move into the hive, the old colony will grow weaker and will be unable to maintain itself.

5. About 4 weeks later, remove the cone. Bees from the new hive will enter the building and transfer the honey to the new hive. The queen in the building often is lost. A new queen must be provided if the bees did not produce one from the brood supplied in the hive. Remaining bees will move to the hive, leaving only the empty wax comb in the building. After the bees have moved completely, close all holes and cracks to prevent another swarm from entering.

Source: Revised by Eric Mussen, Apiculturist, Cooperative Extension, UC Davis, from an original publication by Ward Stanger, Apiculturist Emeritus, Davis.

FIGURE 6-15.

Traps baited with attractants may reduce the numbers of foraging adult wasps in specific locations. Common attractants include raw beef or tunafish.

Managing social wasps in and around buildings requires skills similar to those used for honey bees. A few additional techniques are available; one is to use traps baited with an attractant, with the hopes of reducing the numbers of foraging adults that are pests in specific locations (Figure 6-15). When traps become sufficiently full they can be submerged in water to destroy the wasps. This control method may not work well when yellowjacket wasp populations are high. Poisonous bait will be taken back to the nest and destroy the entire colony. This method is slow and may require time before results can be seen; furthermore, poisoned baits only work well during late summer when other food sources become depleted. Suitable bait includes treated meat or tuna; do not use sugars or syrups as these attract and poison beneficials such as honey bees. Keep all bait materials out of the reach of children and pets.

One way of removing adults from a ground nest is by vacuuming them up as they emerge from the nest opening. If you have never removed a nest in this manner, work with someone who has in order to gain experience. Be sure to wear protective clothing to prevent getting stung (heavy coveralls with leg and sleeve openings sealed with tape, gloves, boots, and a hat with a bee veil). Seal all but one opening before beginning (it may be difficult to locate and seal all openings). Two people are required to successfully destroy a nest: one opens the nest while the other operates the vacuum. Destroy adults picked up in the vacuum as well as larvae removed from the nest by freezing them.

It is also possible to use a quick-acting liquid insecticide to destroy social wasp nests in the ground, wall voids, or other locations. This is done the same way as for control of honey bees. Be sure to seal off all openings except the one used to apply the insecticide. Pour or spray a large quantity of material directly into the nest to prevent any adults from escaping.

Certain insecticide dusts can also be used for control of social wasps. Seal off all but one nest opening, then blow the dust into the nest. This method works well if some of the nest population is away from the area. They will brush against the dust upon return and be killed.

If you must destroy or remove hornet or yellowjacket nests, it is best done at night, when all individuals are in the nest rather than out foraging for food. At night, too, cool temperatures inhibit wasps from flying. When using a flashlight, cover the lens with red cellophane to make the light invisible to the insects.

Umbrella wasps (genus *Polistes*) are beneficial and are generally not aggressive. They have much smaller colonies than the hornets and yellowjackets, usually up to 200 individuals per nest. Nests have a single layer of exposed cells and are found in protected places such as under eaves or roof tiles. Nests can be removed with a pole or stick.

Solitary Wasps

Solitary wasps, of the family Sphecidae, include several species of mud daubers that range in length from ½ inch to over 1¼ inches (Figure 6-16). Some of these wasps are distinctive, having the abdomen separated from the thorax by a long, slender waist or *petiole*. One species, *Sceliphron caementarium*, is black and yellow, and another, *Chalybion californicum*, is dark metallic blue with blue wings.

As implied by their name, female mud dauber wasps construct cells of mud. They provision these with spiders or insects as larval food for their off-

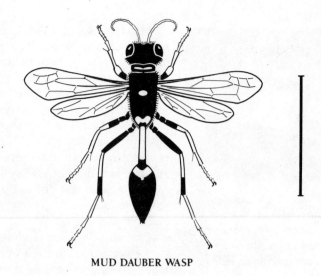

MUD DAUBER WASP

FIGURE 6-16.

Mud dauber wasp (family Sphecidae).

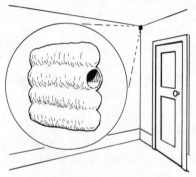

MUD DAUBER'S MUD NEST

FIGURE 6-17.

Female mud dauber wasps construct mud cells which they provision with spiders or insects as food for their young.

spring (Figure 6-17). When the cells is full, a single egg is deposited and the cell is sealed. An adult wasp assembles several cells together in one nest. Nests are located in protected places among trees and rocks in nature, but the wasps also construct them under eaves of buildings and in attics and other out-of-the-way areas.

Mud daubers serve as beneficial insects because they contribute to the control of spiders, including black widow spiders, around buildings and in attics. These wasps are usually few in number and are not aggressive like bees or yellowjackets. Their mud nests may be unsightly and may occasionally attract carpet beetles.

Management Guidelines for Solitary Wasps. Control mud dauber wasps by physically destroying the mud cells and blocking access to attics or other areas inside buildings. Use a spatula, pole, or broom to dislodge mud nests from walls, ceilings, or eaves. In outdoor areas, the nests can also be removed with a high-pressure stream of water. Cells that have an external opening in them are empty—the opening was made by a newly emerged adult. These cells should be removed, however, because they contain food that may attract other household pests such as carpet beetles.

SPIDERS

Spiders are typically seen in buildings, in attics and crawl spaces, under eaves, around windows, and in shrubbery outside of buildings (Figure 6-18). Many different species of spiders may be found; some make webs to snare their prey, and others are hunters that are seen walking across floors or climbing walls in search of food. All spiders are predators that depend mostly on live insects for food. They are highly beneficial because they consume many pest insects such as flies, cockroaches, and mosquitoes. Spiders never feed on plants or any type of nonliving material such as grains or fabrics. They are pests primarily because they leave unattractive, dust-catching webbing in corners of rooms, around windows, and on outside surfaces of buildings and plants.

FIGURE 6-18.

Spiders commonly occur in and around buildings.

Dense webbing filled with dust and insect remains is also found in basements, crawl spaces, garages, and other dark places. This debris attracts other pests such as carpet beetles.

Some spider species are capable of inflicting painful venomous bites, but rarely do. The black widow is the most notorious biting spider, although large garden spiders have occasionally been known to cause injury to people who accidentally come too close. The brown recluse or violin spider is not common in California. Despite spider bites being rare, spiders have a frightful reputation to the point where many people cannot tolerate their presence.

Description, Development, and Habits. Several species of spiders may be encountered in buildings. Each species has unique habits, food preferences, and life spans, although it is possible to generalize on some aspects of their biology.

Most spiders lay eggs into an egg sac constructed of special webbing. Egg sacs usually contain several hundred eggs, and most female spiders construct two or more of these during their adult life. Females of some species carry the egg sac with them attached to the end of their abdomen, under their body, or in their jaws. Other species position the sac in their web and stand guard over it. Generally, female spiders are very protective of their eggs and newly hatched young.

Eggs hatch in 2 to 3 weeks, but young spiderlings of most species remain in the egg sac for several more days. The total time between egg laying and the appearance of young spiderlings outside the egg sac is usually 3 to 4 weeks. Newly hatched spiders resemble adults but are much smaller—usually they are light-colored and have few distinctive markings.

As spiderlings grow, they shed their outer body covering, a process known as molting; shed "skins" are called exuviae. The first molt takes place before spiderlings emerge from the egg sac. Spiders usually undergo six or seven molts before reaching maturity; most species found in and around buildings reach maturity within 6 months. With the onset of the winter season, however, immature spiders do not complete their final molt until spring. Spiders grow larger with each successive molt and in their coloration and markings gradually come to resemble adults.

Adult females are larger and more robust than males and usually live longer. Adult females generally live for 6 or 8 months, although some species, for example large female tarantulas, may live for more than 30 years. As adults, males of all species cease feeding and concentrate their efforts on finding females for mating—their adult lifespan only lasts a few months.

Spiders produce several types of silklike webbing from glands in their abdomens. Webbing is emitted through external structures at the end of the abdomen called spinnerets. Most young spiders disperse from their hatching area by letting out a long strand of webbing, or gossamer, which carries them aloft on air currents. Spiderlings can travel hundreds of miles by "ballooning" in this manner.

One characteristic of some spiders is the web they construct to catch prey. Some are symmetrical orbs constructed between supporting structures, such as those made by garden spiders. Others are irregular jumbles known as cobwebs; these are constructed by such species as the black widow and the long-bodied cellar spider. Other species make sheet or funnel webs to entice and capture prey. Not all spider species construct webs; many roam in search of food or hide among flowers or foliage and are adept at jumping on and overpowering prey.

All spiders inject paralytic venom into their prey through hypodermiclike fangs. Spiders also use their fangs to defend themselves from enemies or danger, and this is the source of spider bites.

Management Guidelines for Spiders. Most spiders found indoors and around the outside of buildings are harmless to people and are often beneficial but are nevertheless controlled because they or their webs are nuisances. Spiders depend on live insects or other small arthropods for food, but they can survive for long periods without food. They are most numerous in locations where insects occur. Because of the innocuous nature of most spiders, control may not be needed. In most situations, control should be nonchemical; reserve pesticide use to destroy black widow or other occasionally harmful species if needed.

Satisfactory management of spiders inside buildings begins by eliminating food sources and by keeping spiders out. Items such as firewood, cut flowers, and nursery plants should be thoroughly inspected for spiders or spider eggs before being brought indoors. Eliminate the spiders' food by keeping flying insects such as flies and mosquitoes out of buildings. Use door and window screens and insect traps, and caulk or fill small cracks and other openings. These techniques also help exclude spiders, although most young spiderlings are so small they can pass through the openings of window screening and small gaps in poorly fitted doors and windows. If food is not available, however, spiders will not survive. Whenever possible, use a vacuum cleaner to remove spiders and webbing from ceilings, corners, and behind and under furniture and appliances. Be sure to remove egg cases.

Black widow spiders (Figure 6-19) usually are found only in dark locations. Look for them behind or under furniture, in closets, basements, attics, crawl spaces, and storage rooms. When controlling poisonous spiders in inaccessible locations, apply an insecticide as an aerosol or fine mist. The spray must actually contact the spider or its webbing to be effective because these spiders are not mobile and do not usually walk over sprayed surfaces; they may even avoid treated areas. Use only pesticides specifically registered for spider control inside buildings.

Apply desiccants (inert dusts or sorptive powders) in basements, attics, wall voids, crawl spaces, store rooms, garages, and other similar areas where spiders are found. Before applying a desiccant, clean the area to remove spiders, eggs, and webbing. Apply dusts to surfaces where spiders attach their webs. Desiccants are effective in providing relatively safe, long-term control as long as they remain dry. A pesticide dust or liquid residual spray can also be used in these areas for control of spiders, although control may not last as long.

Remove spiders from building exteriors by vacuuming, sweeping, or washing with a high-pressure stream of water. Insects are attracted to lights, therefore spiders congregate in these areas too; if possible, relocate exterior lights to attract insects away from buildings or use lights that are less attractive to insects, such as sodium vapor lights or yellow incandescent "bug" bulbs. To eliminate suitable habitats, shrubbery should be trimmed away from buildings, and debris and items stored next to buildings should be removed. Periodic cleaning of building exteriors is usually sufficient to prevent excessive spider problems and eliminate the need for chemical controls. If further treatment is necessary, apply a liquid residual pesticide spray labeled for control of spiders to surfaces where spiders congregate, such as beneath eaves and around windows and doors. Also apply this type of spray around the foundation of the building where spiders must pass to gain access to walls or eaves; this application may help keep spiders out of buildings as well.

FIGURE 6-19.

Black widow spiders are usually found in dark locations inside or around the outside of buildings.

7 Fabric Pests

Fabric pests include insects that feed on natural fibers, synthetics, and animal byproducts. They damage clothing, upholstery, carpeting, draperies, and other fabrics. Some of these pests are able to digest the animal protein keratin and therefore feed on hides, furs, hair, feathers, animal horns, and preserved insects and other museum specimens. Several fabric pests are also important stored-product pests (such as black carpet beetles, silverfish, and firebrats).

Four orders of insects have species considered to be fabric pests: the Coleoptera (carpet beetles); the Lepidoptera (clothes and webbing moths); the Thysanura (silverfish and bristletails); and the Orthoptera (crickets).

CARPET BEETLES

Beetles make up the very large insect order known as Coleoptera. All beetles undergo complete metamorphosis and in the immature stage have several larval instars (stages between molts). They pass through a pupal stage before becoming adults. Adults are winged and many species are good fliers. Adult beetles are distinctive among adult insects because their front pair of wings is modified into hard body coverings known as elytera. When a beetle flies, the elytera are raised to expose the hind wings. Elytera are shiny and brightly colored in some species of beetles; other species have a covering of fine hairs or scales.

Carpet beetles belong to the coleopteran family Dermestidae. In California, three species of carpet beetles cause serious damage to fabrics, carpets, furs, stored foods, and preserved specimens. These insects are pests in warehouses, homes, museums, and other locations where suitable food exists.

Varied Carpet Beetle
Anthrenus verbasci

The adult varied carpet beetle is about $1/10$ inch long and is black with an irregular pattern of white, brown, and dark yellow scales on its elytera. Older adults, however, may have lost the scales which form this pattern, so they appear solid brown or black. Figure 7-1 illustrates the characteristics that distinguish the varied carpet beetle from the furniture carpet beetle, discussed next.

Female beetles search out nests of bees, wasps, birds, and spiders in which to lay their eggs because these contain dead insects, beeswax, pollen, feathers, or other nest debris that can serve as larval food. These nests may be a source of infestation of varied carpet beetles into a building. Once inside, beetles deposit eggs on or near wool carpets and oriental rugs, woolen goods, animal skins, furs, stuffed animals, leather book bindings, feathers, horns, whalebone, hair, silk, dried plant products, and other materials that can be used as larval food.

About 40 eggs are laid by each female; eggs hatch in about 18 days. Larvae pass through 5 to 16 instars depending on temperature, humidity, and food

FIGURE 7-1.

Varied carpet beetle, Anthrenus verbasci.

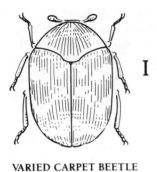

VARIED CARPET BEETLE

I

quality. Mature larvae are about the same length as adults and are covered with dense tufts of hairs which they can extend to form a round plume if disturbed. Larvae have a series of transverse stripes alternating between light and dark brown, and are distinguishable from other carpet beetle larvae because they are broader in the rear and narrower in front. The larval stage may be completed in 220 to 320 days, or may extend for as long as 630 days, depending on environmental conditions and food sources. Pupation takes 10 to 13 days. Adults usually appear in spring or early summer. Adult males live for 2 to 4 weeks and females live for 2 to 6 weeks. When outdoors, adults visit flowers to feed on pollen. Adults found indoors usually are observed near windows in the spring.

Furniture Carpet Beetle
Anthrenus flavipes

Adults of the furniture carpet beetle are slightly larger and more round when viewed from above than the varied carpet beetle (Figure 7-2). Adults have

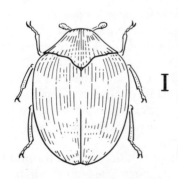

FURNITURE CARPET BEETLE

I

FIGURE 7-2.

Furniture carpet beetle, Anthrenus flavipes.

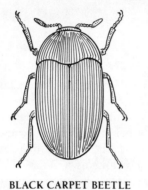

BLACK CARPET BEETLE

FIGURE 7-3.

Black carpet beetle, Attagenus megatoma.

a mottled appearance due to white and dark yellow to orange scales interspersed with black spots on their elytra. They are white on their undersides. They may appear solid black if these scales are rubbed off. Coloration and markings are highly variable.

Females lay a total of about 60 eggs in one to three clutches on surfaces of upholstered furniture, clothing, and in cracks and crevices. Hatching begins in 9 to 16 days. Larvae are white at first but darken to dark red or chestnut brown as they mature. In contrast to larvae of the varied carpet beetle, larvae of the furniture carpet beetle are broadest in front and narrower at the rear. The larval period lasts 70 to 94 days and pupation takes from 14 to 17 days. Adults live 4 to 8 weeks. Feeding habits of larvae are similar to those of the varied carpet beetle.

Black Carpet Beetle
Attagenus megatoma

Larvae and adults of the black carpet beetle (Figure 7-3) are distinctly different from the other carpet beetles described above. In California and other arid areas, the black carpet beetle is a more serious stored-product pest than a fabric pest.

Adult black carpet beetles range from ⅛ to ³⁄₁₆ inch in length. They are shiny black and dark brown with brownish legs. Full-sized larvae may be as long as ⁵⁄₁₆ inch. These range in color from light brown to almost black. Larvae are shiny, smooth, hard, and covered with short, stiff hairs; they resemble larvae of wireworms. Their body tapers toward the rear and terminates in a tuft of long hairs.

It requires about 180 to 360 days for a black carpet beetle to develop from an egg to an adult. The egg stage usually lasts from 6 to 16 days. The larval period takes from 166 to 330 days, and pupation lasts from 8 to 14 days. These insects usually pass through 5 to 11 instars; however, under certain conditions, 20 instars have been observed. Adults live for about 30 to 60 days.

Females produce about 90 eggs, which are usually deposited in dark, protected places.

Management Guidelines for Carpet Beetles. Carpet beetles are among the most difficult building pests to control due to their ability to find food in obscure places and to disperse widely throughout a building. Control success depends on integrating the use of sanitation, exclusion, and, where necessary, insecticides.

Monitor for adult carpet beetles using sticky traps baited with an appropriate pheromone. Several traps located throughout a building can show where beetles are coming from; they are also useful for monitoring the effectiveness of control applications. Check traps once or twice a week. Pheromone traps can also be used to augment other control methods when used to attract adult males in small, confined areas.

Eliminate accumulations of lint, hair, dead insects, and other debris that serves as food for carpet beetles. Destroy any badly infested clothing, rugs, or other items. Bird, rodent, or bee and wasp nests may harbor infestations, as may spider webs with their accumulation of dead insects. Cut flowers brought into a building may harbor adult beetles.

Regular and thorough cleaning of rugs, draperies, upholstered furniture, closets, and other locations where carpet beetles congregate is an important preventive and control technique. Frequent, thorough vacuuming is an effective way of removing food sources as well as carpet beetle eggs, larvae, and adults. Fabrics can be protected by keeping them cleaned, because food and perspiration stains on fabrics attract carpet beetles that feed in the stained areas. Mounted animal specimens, such as museum specimens or trophies, should be regularly cleaned or periodically placed in a freezer for several hours. Stored woolens, linens, and furs should be periodically inspected then aired, brushed, and hung in the light. If infestations are found, launder or dry clean these items before storing to destroy carpet beetle adults, larvae, and eggs. Be sure cleaned items are sealed in a protective plastic bag or other suitable container.

Apply residual insecticides as spot applications. Confine insecticide applications to the edges of floor coverings, under rugs and furniture, the floors and walls of closets, shelving where susceptible fabrics are stored, and in cracks and crevices and other lint-accumulating areas. Use dust formulations, including desiccants (inert dusts and sorptive powders) in attics and wall voids and other inaccessible places. Fumigation may be needed when infestations are extensive, although success can be limited by the ability of the fumigant to penetrate all the areas where carpet beetles hide. Fumigants such as naphthalene can be used in small, tightly closed containers. However, insecticide-impregnated resin strips, labeled for control of carpet beetles on fabrics, are usually more effective in protecting susceptible objects in enclosed containers. These strips slowly release an insecticide vapor providing prolonged protection. Infested furniture or similar objects can be removed from the building and treated in fumigation vaults.

Some insecticides may cause staining or running of fabric dyes, so whenever in doubt, test the chemical on an inconspicuous part of the fabric before making a complete application.

CLOTHES MOTHS

The webbing clothes moth and the casemaking clothes moth are occasional fabric pests in California. Clothes moths belong to the insect order Lepidoptera, family Tineidae. They undergo complete metamorphosis from larvae to pupae then adults. Although many adult moths are attracted to lights, clothes moths are not. They hide when disturbed and adults are rarely seen close to the source of infestation. Larvae of clothes moths spin silken webs, which may be the only sign of the pest's presence.

In years past, sheep that had been treated with chlorinated hydrocarbon insecticides such as endrin, toxaphene, or DDT to protect them against external parasites supplied insect-resistant wool. However, newly produced woolen items are more susceptible to clothes moth infestation because these persistent insecticides are no longer being used as a sheep dip or spray. As a result, there has been an increase in clothes moth problems, requiring other types of protective measures. Heavy reliance on synthetic fibers has helped to reduce the clothes moth problem.

Webbing Clothes Moth
Tineola bisselliella

The webbing clothes moth (Figure 7-4) is the most common fabric moth. Adults are golden colored with reddish golden hairs on top of the head. Wings, with a wingspan of about ½ inch, are fringed with a row of golden hairs. Adult moths are not attracted to lights, but are usually found very close to the source of infestation such as in dark areas of closets.

WEBBING CLOTHES MOTH

FIGURE 7-4.

Webbing clothes moth, Tineola bisselliella.

Females lay an average of 40 to 50 eggs over a period of 2 to 3 weeks and die once egg laying has been completed. Adult males outlive females and continue mating throughout their adult life. Eggs are attached to threads of fabric with an adhesive secretion; they hatch in 4 to 10 days during warm weather. Larvae molt from 5 to 41 times depending on indoor temperatures and type of food available. The larval period lasts from 35 days to 2½ years. Larvae are shiny white with a dark head capsule. They spin webbing as they feed and may partially enclose themselves in a webbing cover or feeding tube. Feeding tubes are usually extended along floor cracks under carpets. Excrement of the webbing clothes moth may contain dyes from the cloth fibers being consumed and thus be the color of the fabric they are infesting; this same color appears as a median streak seen through the outer parts of the larvae.

Pupation lasts from 8 to 10 days in summer, 3 to 4 weeks in winter. Heated buildings enable webbing clothes moths to pass through their life stages more rapidly during winter months.

Larvae feed on wool clothing, carpets, rugs, upholstered furniture, furs, stored wool, animal bristles in brushes, and even wool felts in pianos. Synthetics are also fed on, especially if blended with wool. Larvae may use cotton fibers to make their pupal cases. Damage generally appears in hidden locations such as under collars or cuffs of clothing, in crevices of upholstered furniture, and in areas of carpeting covered by furniture. Fabrics stained by foods, perspiration, or urine are more subject to damage.

Casemaking Clothes Moth
Tinea pellionella

Adults of the casemaking clothes moth are roughly the same size or slightly smaller than the webbing clothes moth but are similar in appearance (Figure 7-5). They can be distinguished from the webbing clothes moth by their wings, which are more brownish, and their forewings, which are dimly spotted with a darker color. Also, hairs on the head are lighter colored than those of the webbing clothes moth. Larvae of both species are nearly identical, except that larvae of the casemaking clothes moth always carry a silken case with them as they feed. They never leave this silken tube, but enlarge it as they grow. They feed from either end and retreat into it when disturbed. This case takes on the coloration of the fabric that larvae feed on. Pupation also takes place inside the case.

In its food preferences and biological development, the casemaking clothes moth is very similar to the webbing clothes moth.

Management Guidelines for Clothes Moths. Control of clothes moths depends on preventing infestation, protecting fabrics, and selectively using insecticides when necessary. If humidity can be kept low inside buildings, this will create an environment that is not suitable for clothes moth development. Building construction that is free of many tiny cracks and crevices also contributes to fewer clothes moth problems.

Regular, thorough cleaning of susceptible clothing, carpets, closets, and storage areas is an important factor in clothes moth control. Strong vacuums should be used to remove eggs and larvae. Clothing and other fabrics should be periodically shaken and brushed to remove these insects or their eggs; pay special attention to seams, collars, and cuffs. To keep from attracting moths, launder or dry clean soiled fabrics before they are stored or hung in a closet. Whenever possible, store garments, blankets, linens, and rugs in tightly sealed boxes or containers. Cold storage at temperatures between 40° F and 42° F

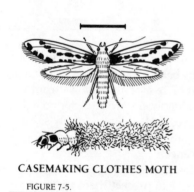

CASEMAKING CLOTHES MOTH

FIGURE 7-5.

Casemaking clothes moth, Tinea pellionella.

can further protect expensive clothing and furs from moth damage. Cold storage is also effective in killing moths if they are exposed to rapid changes of temperature, for example a sudden change from 50° F to 18° F, before storage at 40° F to 42° F.

Pyrethrins insecticides provide quick knockdown of clothes moths, and most can be sprayed directly on fabrics if needed (in situations where fabrics cannot be laundered or dry cleaned). Some pyrethrins insecticides do not leave persistent toxic residues, and are especially suitable for clothes moth control. Use a residual spray along baseboards, margins of carpets, in closets, and in storage areas. Also spray under furniture and other areas where moths occur. Before treating any fabric with an insecticide, test a small, inconspicuous part of the fabric to be certain the spray will not cause staining or running of dyes.

SILVERFISH AND FIREBRATS

Silverfish and firebrats make up the insect order Thysanura, which are among the most primitive insects. There are about 13 species of silverfish and firebrats in the United States. These small, wingless pests (Figure 7-6) do not undergo complete metamorphosis; hatchlings look like adults but are smaller. All life stages have similar feeding habits. Immature forms may molt as many as 50 times before becoming adults. Silverfish and firebrats continue to molt during the adult stage. These insects are long-lived, taking up to 2 years to reach maturity and then continuing to live for several years as adults. Under optimum conditions, the immature stage lasts for 2 to 3 months.

Adults range from ¼ to ¾ inch in length, depending on the species. Silverfish are tapered in the back, giving rise to their fishlike appearance. Most are silver colored; firebrats are gray with darker markings. Both silverfish and firebrats have long antennae and three long bristles, known as cerci, arising from the tip of the abdomen—because of this they are sometimes called bristletails. Females lay approximately 100 eggs during a lifetime; eggs are deposited as small batches, usually in cracks or obscure places. Eggs require 2 to 8 weeks to hatch.

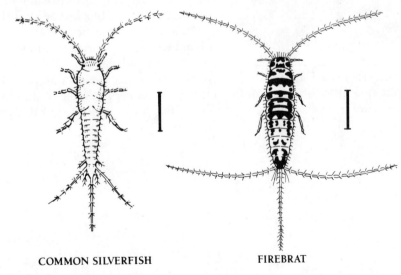

COMMON SILVERFISH   **FIREBRAT**

FIGURE 7-6.

Firebrat, Thermobia domestica *(right and in photo) and silverfish,* Lepisma saccharina *(left).*

Silverfish and firebrats gain entry into buildings through openings in foundations or around pipes or wires passing through walls. They can also be carried into buildings in boxes, books, papers, or other items brought from infested areas. These insects are attracted to buildings and survive if the area has a warm, moist environment and suitable food. They live in most locations in a building including attics, basements, and wall voids. Firebrats require warmer areas than silverfish and can tolerate drier conditions. Both silverfish and firebrats are nocturnal and are not attracted to light—they are rarely seen in well-lighted locations.

Silverfish and firebrats feed on fabrics such as linen, rayon, and cotton. They are attracted to starched fabrics and also feed on paper, paper sizing, book bindings, and dead animals. They feed on any type of human food, but appear to be especially attracted to flour and starches; they may also be found in breakfast cereals. They do not feed on wool, hair, or other animal fibers but may damage some synthetics. Silverfish and firebrats are voracious eaters, but are also capable of going without food for long periods. Besides damaging objects by feeding, silverfish and firebrats leave yellow stains and dark-colored feces on items they have contacted.

Management Guidelines for Silverfish and Firebrats. Due to their nocturnal habits, silverfish and firebrats are difficult to see. If possible, make observations or surveys of silverfish or firebrats during the night, using a flashlight. They may also be monitored with sticky traps. These insects may go unnoticed until populations get large or damage becomes severe. Control may be difficult because it is hard to locate the sources of infestation.

Keep silverfish and firebrats from entering buildings by caulking or otherwise closing openings from the outside. Caulk cracks and fill other openings inside the building to eliminate hiding places (Figure 7-7). Moisture attracts these insects, so it is important to repair leaking pipes and drains and insulate water pipes to prevent water condensation. Wherever possible, eliminate sources of food; store flour, cereals, and similar items in tightly sealed containers.

Chemical control methods for firebrats and silverfish are similar to those used for German cockroaches. Insecticides may be sprayed around building foundations; desiccant dusts may be applied to attics, crawl spaces, and voids in walls and beneath cabinets. Insecticides used in this manner create barriers that keep silverfish and firebrats out of building interiors when openings cannot be blocked or located.

Apply a liquid insecticide having residual activity to locations where silverfish or firebrats are most concentrated. Inject the spray into cracks and crevices. Dust formulations may be preferred in dry areas where visible residues are not objectionable. Blow dusts into wall voids, attics, and into cracks and crevices. Do not apply liquid or dust formulations to books or papers or other objects that might be stained or that people come into close contact with.

FIGURE 7-7.

To help eliminate hiding places for firebrats or silverfish, caulk cracks and fill other small openings.

CRICKETS

Crickets are sometimes nuisances in buildings and they may also damage fabrics or other materials. They are especially destructive to silks and woolens. They are also attracted to perspiration and other stains on clothing and fabrics. Occasionally crickets invade a structure in large numbers. They are often attracted to lights around a building at night. Besides the damage they may cause, they produce a chirping sound which may, after a period of time, become annoying to building inhabitants.

Crickets belong to the insect order Orthoptera and are related to grasshoppers. These insects do not undergo a complete metamorphosis, therefore the young resemble adults except they do not have functional wings. Young and adults both have similar feeding habits.

The most common crickets to invade buildings include the house cricket, *Acheta domesticus,* and the field cricket, *Gryllus* spp. (Figure 7-8), which are very similar in appearance. A more recent cricket pest, introduced from Arizona, is the Indian house cricket, *Gryllodes suppilean.*

House cricket adults range in length between ½ and ¾ inch. They may be light yellowish brown, with three dark bands on the head, or solid shiny black. This species has long, slender antennae. The field cricket is slightly larger, up to 1 inch in length, and usually brown or black. Females of both species have a long, thin ovipositor projecting from the tip of the abdomen.

Management Guidelines for Crickets. The key to managing crickets in buildings is exclusion. Cracks and other openings from the outside that provide access to the building should be sealed. Caulk or otherwise seal cracks and crevices inside the building that provide hiding places. Behind or under heavy furniture and appliances or in other inaccessible areas, it may be possible to remove crickets using a strong vacuum cleaner. Weeds and debris

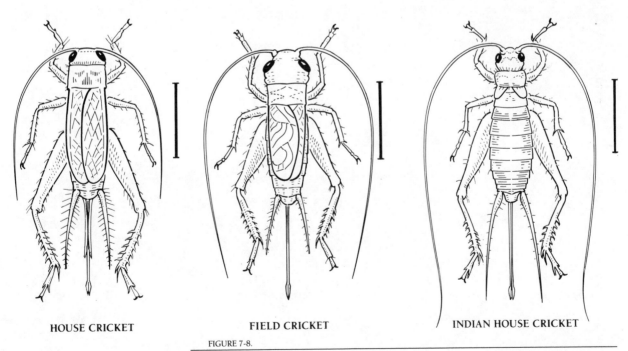

HOUSE CRICKET   **FIELD CRICKET**   **INDIAN HOUSE CRICKET**

FIGURE 7-8.

House cricket, Acheta domesticus, *field cricket,* Gryllus *spp., and Indian house cricket,* Gryllodes suppilean.

around the outside of the building should be removed to eliminate attractive habitats. Change outside lighting to sodium vapor lights or yellow incandescent lights that are less attractive to crickets (and other insects). Garbage and other refuse that serves as food should be stored in containers with tight lids and elevated off the ground on platforms or bricks.

Insecticides should be used only when exclusion and sanitation cannot accomplish control quickly enough to stop the damage within a reasonable time. Use liquid sprays of an insecticide registered for use indoors as a spot spray in cracks and crevices and other areas where crickets may hide. Sorptive powders may also be blown into inaccessible areas. Apply liquid sprays around the perimeter of the building or in other outdoor areas if crickets cannot be controlled through sanitation. Avoid using outdoor spray materials in indoor areas unless the label states that this is permissible. Insecticide-impregnated baits or granular formulations of certain materials may also be used outdoors around buildings for cricket control. Granules are suitable in lawns and other areas subject to moisture or frequent watering. Avoid the use of baits or granules if children or pets can gain access to them.

Cricket infestations are usually seasonal. Most often problems occur during the fall as evenings become cooler and the insects seek buildings for warmth and shelter. Because of this, applications of long-residual insecticides are not usually needed indoors for adequate control.

8 Stored-Product Pests

Many species of pests infest and damage stored cereals, grains, nuts, dried fruit, and other food products. These include birds, rodents, fungi and other microorganisms, mites, and certain insects such as weevils, beetles, moths, silverfish, and firebrats. Stored-product pests are widespread and cause serious economic losses to grain producers, food processors, and consumers. They attack stored products on farms and in processing plants, warehouses, grocery stores, restaurants, homes, and virtually any other location where food is stored or prepared.

Contamination of food products by pests or pest excrement cannot be tolerated, even at low levels. Contaminated food may contain disease-causing organisms or toxins that can cause human illness. Some types of pest infestation destroy or damage the food's nutritional value or change its physical properties. Even if the contaminated food remains healthful, it lacks aesthetic appeal. Some contaminated grains may require special cleaning and screening to remove the pest and its damage; nuts and dried fruit may need to be hand-sorted. The added expense of these processes increases production costs of the food.

To control losses due to stored-product pests, (1) use management methods that prevent pest infestation, (2) eradicate existing infestations, and (3) stop the spread of the pests or contamination to other food items. Establish an integrated approach that includes periodic inspection and monitoring, sanitation, exclusion, and appropriate chemical and nonchemical controls. Use mechanical techniques such as aerating the stored products for moisture control, controlling storage temperature to reduce moisture condensation or uptake and to prevent development of insects, and rotating or turning the stored products to stop localized pest outbreaks. Never store pest-free items near infested products or in contaminated or infested containers or buildings.

Use pesticides as one of your management tools when needed to stop the buildup of pests and to supplement other control methods. Follow pesticide label directions carefully and be certain that application equipment is properly calibrated.

BIRDS

Birds can consume large quantities of grain and other items, and they may also contaminate stored food with their feces and feathers. Bird feces may contain salmonella bacteria and fungal spores that can produce serious intestinal poisoning of people.

The most important way to prevent bird damage is to exclude them from storage areas. Areas where birds are most apt to be a problem are warehouses with large doors that are kept open. If doors cannot be kept closed, install nets or strips of plastic or fabric (Figure 8-1) at the entrances. These barriers enable people and vehicles to pass through freely but keep birds out. In all storage facilities, seal cracks and openings that are large enough for birds to pass through. Close off vents and other high-level openings with wire screen

FIGURE 8-1.

Use nets or plastic strips over large doors to keep birds out.

having a mesh of ¼ inch or smaller. Remove or modify ledges that serve as roosting sites or install nets or other barriers to keep birds from roosting in or on the storage facility. Other attractive nearby roosting sites, such as large trees, may also need to be eliminated.

Maintain good sanitation practices so birds are not attracted to storage areas. Clean up grains or other items spilled during loading, transfer, or handling. Be sure that conveyors, railings, ledges, and other parts of the storage facility are kept clean and free of food residues. Dispose of spoiled or contaminated products in covered containers and remove these promptly from the storage area.

With persistence, certain species such as pigeons can be trapped. Trapped birds are generally released in an area distant from where they were caught.

Avicides are not generally effective in controlling birds when there is an abundance of other food in the area. If avicides are used, place them in locations where there is no risk of contaminating any stored food products. Whenever possible, use materials that repel pest birds rather than killing them.

Trapping, repelling, or poisoning pest birds requires considerable experience and expertise. Permits may be required from the California Department of Fish and Game for some species. Extreme care is required to prevent injury to protected nontarget species.

RODENTS

Rodents such as rats and mice are troublesome pests of stored food. Rodents can chew through wood and other materials to get to food sources. They are good climbers and can squeeze through small openings. Rats and mice can rapidly build up their populations and then consume or contaminate large quantities of stored food. They contaminate stored products and storage facilities with their urine, feces, and hair, and they damage cloth, plastic, and paper bags or cardboard boxes used to package stored products. Rodents within a storage facility may also chew on electrical wiring and cause serious fire hazards or malfunction of equipment.

Exclusion. The most important control method for rodents is rodentproofing. To exclude rodents from storage areas, seal openings with heavy gauge sheet metal, heavy wire screen with a mesh of ¼ inch or less, or concrete with heavy wire screening embedded in it. Attach metal plates to the bottoms of doors, as shown in Figure 8-2, to reduce the gap to ¼ inch or less and prevent rodents from gaining entry. Modify foundations of buildings with concrete or metal barriers to stop rodents from digging their way in. Eliminate dead spaces inside the storage area to restrict areas where rodents may hide. Dead spaces include double walls, false ceilings, enclosed staircases, boxed plumbing, and voids under cabinets.

Sanitation. Sanitation is important in preventing rodent buildup. Spilled grains and other food items around the periphery of a building attract rodents and encourage them to nest nearby. Be sure all spills are cleaned up quickly and placed in rodentproof containers or promptly destroyed. Sanitation must also include keeping all storage areas and adjacent spaces well lighted, clean, and orderly. Eliminate weeds, shrubs, and vines that provide shelter and hiding places for rodents. Rodent activity can be more quickly spotted in clean, orderly areas, enabling you to begin control measures early.

Trapping, Baiting, and Fumigation. Rodents infesting a storage facility are controlled by trapping, use of poison baits (rodenticides), fumigation, or combinations of these methods. See Chapter 10 for complete information on ways to control rodents with traps and rodenticides.

When controlling rodents in food storage areas, consider the following points. (1) Trapping requires daily checking for trapped animals and servicing of the equipment—if traps are baited, the bait must be kept fresh by replacing it periodically. (2) Poisonous baits must be kept fresh to be attractive, therefore bait stations need to be checked and refilled frequently. If baits are the multiple-feeding anticoagulant type, rodents must feed on them continually over a period of several days. (3) Once started, bait stations must not be allowed to become empty, otherwise rodents may recover from the toxic effects. (4) Use of rodenticides such as poison baits within storage facilities creates the risk of product contamination and may not be allowed in some situations. (5) Baits may not be very effective as long as the rodents have access to the stored food product. (6) Poisoned animals may wander off and die, making them difficult to locate. Dead animals create smells and attract insects such as flies. (7) Fumigation may leave dead animals in inaccessible places.

When using rodenticides for control of rats or mice inside or around a food storage facility, it is very important to identify the rodent species involved. You need this identification to understand the habits of the rodent so you can select the right rodenticide and use it properly. Mice, for instance, tend to restrict their activities to a small area, probably no more than 30 feet from their nest, and never move beyond this area unless food or shelter is eliminated. Bait placed only a few feet away from a mouse nest will have no effect if the mouse never travels near it. Different species of rodents may inhabit different levels of a storage structure, or different colonies of the same species may even be at different levels. An effective rodenticide or trapping program requires locating all of the rodent colonies and placing control agents within the reach of each colony.

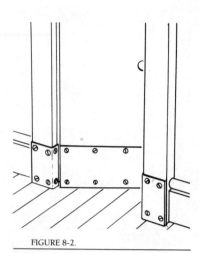

FIGURE 8-2.

Attach metal plates to bottoms of doors to reduce the gap and prevent rodents from chewing through and gaining entrance into buildings.

FUNGI AND OTHER MICROORGANISMS

Many microorganisms may attack and damage stored food products, including bacteria, protozoa, slime molds, yeasts, and filamentous fungi. A large number of these microorganisms require free water to grow and reproduce, therefore these are only problems if the stored products become wet or are wet when they are put into storage. The most serious problem of stored grains and other products, however, comes from filamentous fungi adapted to conditions without free moisture. Damage caused by these fungi includes reduced germination of grain seeds, discoloration of grains and other products, microbiological heating of the stored material, caking, decay, and musty odors. Some fungi produce toxic materials that contaminate stored food products and can cause poisoning if ingested; the most serious of these are the aflatoxins produced by the fungi *Aspergillus flavus* and *Aspergillus parasiticus*.

Many suitable conditions promote fungal development in stored foods. These include high moisture, low temperature, presence of insects or mites, damage to the grain or other stored product, degree of invasion by fungi before items are put into storage, and the amount of foreign material present with the stored product. The length of time items are in storage and the amount and type of air circulation available in the storage area also influence fungal development.

Several things should be done to reduce problems with microorganisms in stored foods. Controlling moisture is very important. The length of time items are to be stored influences the amount of moisture that must be removed before storing. For example, grains being held for long-term storage (greater than 2 years) usually must have no more than a 13.5% moisture content. On the other hand, grains may usually be stored for 4 or 5 months at moisture levels of 18% without developing fungus problems. Differences in temperature between the stored product and the surrounding area may cause condensation of water vapor and produce wet spots, favoring fungal growth. To control temperature and condensation in storage containers, provide air circulation or occasionally turn the material or transfer it from one container to another.

Bulk grains usually contain debris, dust, and broken grain particles; these items are known as fines. If a bulk container is filled by using conveyor belts or augers, fines will accumulate in one area near the spout outlet. A concentration of fines in the stored product impairs air circulation in that area and may promote localized fungal development. Fines are also attractive to certain stored-product insects and may increase insect damage. To prevent fines from accumulating in one area, keep the spout moving while filling the storage container to distribute them evenly throughout the bulk of the stored material.

Some stored-product insects carry in fungal spores on their bodies that may infest stored food items. Moreover, insect feeding damage makes some items more susceptible to feeding by other insects and invasion by fungi. The presence of large accumulations of insects may alter the temperature and moisture content of a stored product and may provide more ideal conditions for fungal development. Therefore, controlling stored-product insects can help reduce fungal problems.

INSECTS

Stored-product insects are small and often difficult to detect. Eggs or larvae commonly pass unnoticed from one part of the food handling system to the next. These are important economic pests that contaminate stored food with their excrement, cast skins, dead bodies, and webbing. They consume or damage large quantities of food, and in damaging packaging materials they cause indirect food damage and further economic loss.

Several species of beetles, weevils, and moths are common stored-product insects. Descriptions of some of these are given below. Management guidelines for these insects are included together, following the descriptions, as the principles of control are the same.

Beetles

Sawtoothed Grain Beetle
Oryzaephilus surinamensis

Merchant Grain Beetle
Oryzaephilus mercator

The sawtoothed grain beetle and the merchant grain beetle are similar in appearance and easy to confuse (Figure 8-3). Adults are about 1/10 inch long and reddish brown to dark brown. Lateral margins of the thorax contain six sawtoothed projections on each side. These are long, narrow beetles with characteristic flattened bodies, giving them access to small cracks and crevices. Both species have well-developed wings, but the sawtoothed grain beetle has not been observed to fly. Adults of both species are usually seen running rapidly over stored food. Larvae have brown heads and their bodies are yellowish, elongate, and segmented, with three pairs of legs. They crawl about actively during feeding.

Adult females lay between 45 and 285 eggs singly or in small batches in or around suitable larval food sources. Eggs hatch in about 8 days. Larvae pass through two to four instars over an average of 37 days, and pupation takes

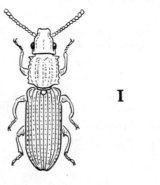 

SAWTOOTHED GRAIN BEETLE   MERCHANT GRAIN BEETLE

FIGURE 8-3.

Sawtoothed grain beetle, Oryzaephilus surinamensis *(left), and merchant grain beetle,* Oryzaephilus mercator (right).

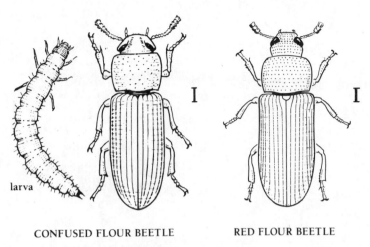

CONFUSED FLOUR BEETLE   **RED FLOUR BEETLE**

FIGURE 8-4.

Confused flour beetle, Tribolium confusum *(left), and red flour beetle,* Tribolium castaneum *(right).*

another 6 days. Temperature and humidity affect the development time and the number of larval instars.

Sawtoothed grain beetle larvae feed on items such as rice, wheat, cereals, and nutmeats. These insects probably cannot attack whole undamaged grains, so may be associated with other pests in whole grains and feed on the kernels that have been damaged by these other pests. The merchant grain beetle is not a major pest of grains or cereals, but prefers seeds and nuts.

Confused Flour Beetle
Tribolium confusum

Red Flour Beetle
Tribolium castaneum

The confused flour beetle and the red flour beetle are the most common and serious pests of flour, cereal, and broken grains (Figure 8-4). They are closely related, similar in appearance, and often occur together. Flour beetles are members of the large coleopteran family Tenebrionidae, commonly known as the darkling beetles. They emit a foul-smelling gaseous secretion when disturbed. Adults are about ⅛ inch long, flattened, and shiny reddish brown. Antennae of the confused flour beetle terminate in four segments that gradually enlarge to form a clublike shape, whereas antennae of the red flour beetle abruptly terminate in three larger, clublike segments (Figure 8-5).

Adult flour beetles live for up to 2 years. Females produce 400 to 500 eggs in their lifetime, laying 2 or 3 per day; eggs hatch in 5 to 12 days. Larvae pass through 5 to 18 instars, typically 7 or 8, over a period ranging from 1 to 4 months. Larvae are slender and wirelike, whitish colored with yellow tinges. They are distinguished from other stored-product insect larvae by the prominent two-pointed termination of the last body segment (Figure 8-6).

Like grain beetles, flour beetles usually do not attack whole grains. They feed on damaged grains, flour, cereals, and other stored products. Their small size provides them access to closed containers that would normally be insect-proof. Adult beetles run quickly when disturbed. In addition to feeding damage, they produce secretions that contaminate the material they feed on, giving it a disagreeable odor and taste.

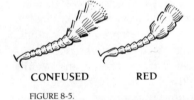

CONFUSED   RED

FIGURE 8-5.

Antennae of the confused flour beetle terminate in four segments that gradually enlarge to form a clublike shape. Antennae of the red flour beetle abruptly terminate in three larger, clublike segments.

FIGURE 8-6.

Larvae of flour beetles have a prominent, two-pointed termination of the last body segment.

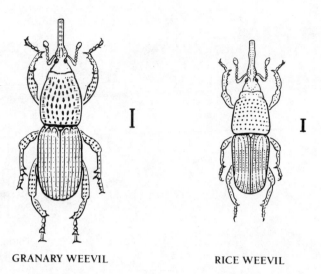

GRANARY WEEVIL   **RICE WEEVIL**

FIGURE 8-7.

Granary weevil, Sitophilus granarius *(left), and rice weevil,* Sitophilus oryzae *(right).*

Granary Weevil
Sitophilus granarius

Rice Weevil
Sitophilus oryzae

Weevils are distinguished from other beetles by the slender elongation of their head, a feature responsible for the common name of snout beetles. Two weevils, the granary weevil and rice weevil, are serious grain pests (Figure 8-7).

Several features distinguish the granary weevil from the rice weevil. The granary weevil is about ⅛ inch long and shiny dark brown or black. The top central area of its thorax is covered with elongated depressions or punctures. Adults have nonfunctional, vestigial wings. By contrast, the rice weevil is a good flyer and is slightly smaller, reddish brown to black, and usually has four reddish or yellowish spots on the elytra. The top central area of the thorax of the rice weevil is covered with round punctures.

Both species bore holes into grain kernels to deposit their eggs. Larvae feed and pupate inside kernels and also feed on caked flour and tightly compressed cereals. Granary weevils have become adapted to living entirely in stored grains and never forage in the wild for food, hence their lack of wings. Rice weevils, however, fly to fields and infest grains such as corn, rice, and wheat. After harvest, infested grain mixed with clean grain causes widespread contamination during storage.

Females lay approximately 200 to 300 eggs during their life (rice weevils produce more eggs than granary weevils). Larvae of both species pass through four larval instars over a period of 3 to 5 weeks and usually have four generations per year. Adults of the granary weevil live from 7 to 8 months when abundant food is available. Adults of the rice weevil live 3 to 6 months.

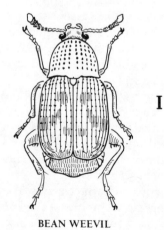

BEAN WEEVIL

FIGURE 8-8.

Bean weevil, Acanthoscelides obtectus.

Bean Weevil
Acanthoscelides obtectus

Bean weevils (Figure 8-8) are not weevils but belong to the seed beetle family Bruchidae. Larvae bore into seeds where they feed and pupate. After pupation, adults bore an emergence hole in the seed. Often more than one larva infest a single seed. Bean weevils are about ⅛ inch long and are light olive brown with darker brown and gray markings and reddish legs. Eggs are laid on pods of legumes such as beans, peas, and lentils in the field or on the surface of stored legumes. Grains, cereals, and other stored food products are not infested by bean weevils. Infestation of stored legumes can easily occur from harvested products being brought in from the field.

Females lay about 75 eggs during their lifetime; these are deposited singly on or near host seeds and hatch after 5 to 20 days. Larvae feed for 4 to 6 weeks before pupating. Adults hibernate during cool weather in the winter, but if temperatures rise, they emerge and females begin egg laying again.

Cigarette Beetle
Lasioderma serricorne

Drugstore Beetle
Stegobium paniceum

Cigarette and drugstore beetles (Figure 8-9) are members of the Anobiidae family, which also includes deathwatch beetles. Adults can be distinguished by their humped appearance due to the head and prothorax being bent downward. The cigarette beetle is reddish yellow to brownish red. Adults are about 1/8 inch long and have the distinctive humped appearance characteristic of this group. Females produce about 30 eggs over a 3-week period; these usually hatch within 1 week. Eggs are attached to sources of larval food such as tobacco, rice, raisins, grains, pepper, and many other stored products. Larvae are curved, plump, and hairy; they are yellowish with a light brown head. The larval stage lasts from 5 to 10 weeks and three to six broods are produced in a year.

Adults of the drugstore beetle are almost the same size as the cigarette beetle. They are reddish brown and can be distinguished from cigarette beetles

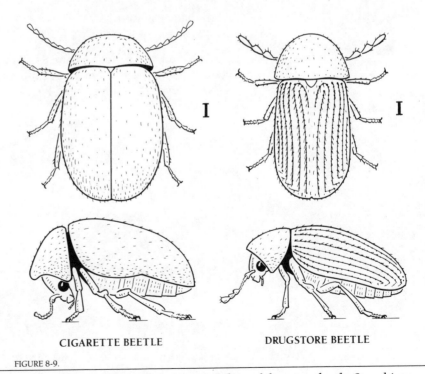

CIGARETTE BEETLE   DRUGSTORE BEETLE

FIGURE 8-9.

Cigarette beetle, Lasioderma serricorne *(left), and drugstore beetle,* Stegobium paniceum *(right).*

by the longitudinal striations, or ridges, on their elytra. They are also less humped.

Drugstore beetles usually have one to four generations per year. They complete a life cycle in about 2 months. Larvae, which resemble cigarette beetle larvae, feed on practically every type of stored product as well as spices, drugs, books, and wood. They can survive on items with low food value because of yeastlike organisms in their digestive systems that produce some essential vitamins.

Black Carpet Beetle
Attagenus megatoma

The black carpet beetle is described in the previous chapter (page 130) as a fabric pest.

This insect is widespread and feeds on a large variety of dried foods including beans, peas, corn, wheat, rice, and many types of seeds.

Moths

Moths belong to the insect order Lepidoptera. Larvae of moths infesting stored food products may be confused with beetle or weevil larvae because of their wormlike shape. Unlike beetles and weevils, the larval stage of the moth is the only stage that causes damage. A telltale sign of infestation is the appearance of small to medium-sized moths in food containers and packaging, or flying around or clinging to walls in a room or storage area.

Indianmeal Moth
Plodia interpunctella

The Indianmeal moth (Figure 8-10) is the most common pest of coarsely ground flours (such as whole wheat flour) and cornmeal. It is widespread in grocery stores, warehouses, and kitchens. The Indianmeal moth also infests

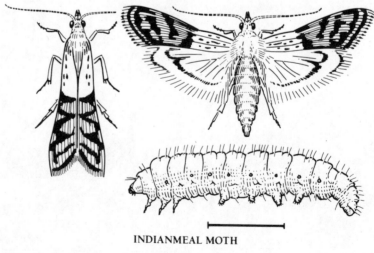

INDIANMEAL MOTH

FIGURE 8-10.

Indianmeal moth, Plodia interpunctella.

shelled or ear corn, broken grains, dried fruit, seeds, peas and beans, crackers, biscuits, nuts, powdered milk, chocolate, candy, red peppers, dry dog food, and other commodities. Unlike weevils and other beetle larvae, Indianmeal moths spin large amounts of webbing, further contaminating food products.

Adults of this moth have a wingspan of about ¾ inch. Wings are pale gray with the outer two-thirds of the forewing colored reddish brown and having a coppery luster.

Egg laying usually begins in April. Females lay eggs at night, either in masses or singly, and produce 200 to 400 eggs. Larvae are a dirty white color but may take on different hues depending on the food being ingested. The larva's head and prothoracic shield are brown. Pupation takes place in a silken cocoon. The larval period varies greatly between 1 and 10 months, depending on environmental conditions and available food.

The normal complete life cycle of this pest takes about 6 to 8 weeks.

Almond Moth
Cadra cautella

The almond moth (Figure 8-11) is a pest of lesser importance than the Indianmeal moth. However, it is capable of causing considerable damage to cereals, dried fruits, flour, grain, seeds, and shelled nuts. Adults are slightly smaller than the Indianmeal moth, having a wingspan of about ⅝ inch. They are a mottled gray color and may have a fawn-colored pattern on the forewing. Larvae are dirty white tinged with brown or purple dots, giving them a striped appearance. They leave matted webbing as they feed.

Females lay an average of 100 eggs, which hatch in about a week. The larval period usually continues for 2 months.

Management Guidelines for Stored-Product Insects

Stored-product insects are tiny and difficult to detect in bulk or packaged food products. Therefore, they can be freely transported from processing plants to warehouses to grocery stores to restaurants and household and

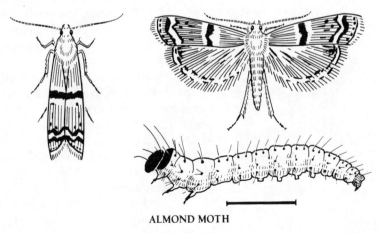

ALMOND MOTH

FIGURE 8-11.

Almond moth, Cadra cautella.

institutional kitchens. Even under the most carefully controlled conditions, it is probable that some of these pests, in egg, larval, or adult forms, will pass from one level of the food handling system to another. Eradication at any one level is virtually impossible due to the size and complexity of the food distribution industry. Once an infestation occurs in one commodity, it can quickly spread to others unless appropriate and timely control measures are taken. Each entity in the complex maze of food distribution, from the producer on down to the ultimate buyer, must assume a role in the management of stored-product insects.

Detection and control methods for stored-product insects have to be ongoing, not sporadic. Management relies on inspection and monitoring to detect and identify pests, followed by an integrated program of control that includes sanitation practices, exclusion techniques, habitat modification, and the careful use of insecticides.

Beetle or moth infestation of a box of cereal or bag of flour in a home is an annoyance. The infestation may result in the loss of the cost of the product and perhaps spread of the pest to other similar products stored in the pantry or cupboard. Control can be as simple as throwing away the infested materials (or returning them to the grocery store for a refund) and storing uncontaminated food products in insectproof containers.

Similar infestations occurring in grocery stores, warehouses, or packaging and processing plants can result in considerable loss of investment and revenue. Pest control efforts, therefore, should be proportional to the potential for loss. Major efforts involving sanitation practices, exclusion techniques, habitat modifications, and insecticide applications are usually required to eliminate damage. Early detection simplifies the management program, reduces control costs, and prevents extensive damage to stored food. Monitoring is used to detect, locate, and identify pests, determine the proper time to apply control techniques, and evaluate the success of the management program.

Inspection and Detection

Inspection and pest detection are necessary parts of a stored-product pest management program. These processes provide information about the pest. They are also important ways to evaluate the control methods used and to follow up afterward to monitor for reinfestation.

Make a complete and thorough inspection of the premises to locate potential sources of infestation. Carefully examine stored food items such as grains, dried fruit, flour, and seeds; dog food may be a major source of infestation. Check outdoor areas surrounding buildings, as some stored-product insects are attracted to certain types of flowers and shrubs, and may also be attracted to outdoor lighting.

Use pheromone traps inside a building or structure to monitor pest activity; pheromones are available for most of the insects that cause damage to stored food (see information on page 10.) Traps utilizing mating pheromones generally catch individuals of one sex, usually males. For other stored-product insect species, traps containing aggregating pheromones are available that attract both sexes. The attractiveness of monitoring systems for some insect species is enhanced by incorporating food attractants with pheromones; food attractants can lure larvae and adults of both sexes. With some species of stored-product insects, food attractants are used alone.

When using pheromones or food attractants for monitoring, place one trap per 250 to 500 square feet of storage space (see Figure 2-3 on page 11). For flying insects, locate traps near storage containers; put them inside containers for insects that do not normally fly.

Sometimes the use of more than one type of pheromone in an enclosed area may prevent target insects from efficiently locating traps. Before installing traps for other species of insects in an area where one type of pheromone trap is already being used, check with the manufacturer or supplier to determine if such a combination will be effective.

Flying insects locate pheromone traps by following a trail of pheromone scent upwind, detecting its increasing concentration in the air. Enclosed areas where traps are located therefore should have some air movement so the atmosphere does not become saturated with pheromone. Keep traps away from bright lights that may repel target insects.

Check traps on a regular basis—daily if there is a low tolerance to stored-product insects on the commodity, or weekly under normal conditions. At each inspection, record the number of pest insects caught and remove them from the traps. Replace pheromones according to instructions supplied by the manufacturer. Change sticky parts of the traps whenever they become so coated with debris that they are ineffective.

Pheromones or attractants can sometimes be used in traps for control of stored-product insects. Trapping may be a preferable control method over insecticide use because foods are not exposed to insecticide residues. Locate traps close to the source of infestation for maximum control and increase the density of traps to about one to each 25 to 50 square feet of storage space.

For stored bulk grains, use pheromones with specially designed probes positioned at different levels inside storage bins. Check probes periodically for the presence of insect pests and use catch data to locate areas of infestation. Monitoring in this manner should also be used to evaluate the effectiveness of other control measures.

Exclusion

Prevent insect entry into the storage facility by inspecting grains, cereals, flour, and other bulk and packaged products as they arrive. Check packages for holes, webbing, insect frass, eggs, living insects, and insect parts. Even new, unused packaging material, such as cardboard, may be a source of insects. Immediately return infested materials to the supplier or destroy them. Never store infested materials in the facility unless they can be enclosed in a tight container or can be refrigerated. Prevent contamination of flour, grains, cereals, or dried fruit by keeping it in insectproof containers. Opened bags or boxes must be resealed securely or their contents transferred to sealable containers. Promptly remove empty boxes and bags from the building.

Keep insects out of buildings by using screens over door and window openings. Close off all other openings with wire screening or caulking. If it is not possible to exclude pests from the entire building, make sure the storage area is protected. Locate and close rodent holes as stored-product insects can enter through these openings. If rodent baits are being used in the area, check the baits for insect infestation; even stored or unused bait may harbor insects. To keep from attracting insects into buildings, locate outdoor lighting away from doorways. Use sodium vapor lights rather than mercury vapor lights for outdoor lighting around warehouses and grocery stores as insects are less attracted to yellow light.

Sanitation

Sanitation is a critical part of controlling stored-product insects in homes, grocery stores, warehouses, and processing facilities. Clean up spilled materials to eliminate food sources for pests. Seal cracks in shelves and bulk food containers to eliminate places where pests can hide and to keep grains, flour, or other food from accumulating. Keep storage shelves far enough away from walls to leave room for cleaning. Raise shelving in warehouses and other storage areas off the floor to make cleaning underneath possible. Areas where susceptible items are stored should be well lighted for ease in cleaning and spotting pest infestations; moths may be easier to detect during evening hours when they are active. Conveyors, augers, and food processing machinery must be thoroughly cleaned on a regular basis to prevent them from being potential pest harborages.

Modifying the Environment

Manipulating storage temperatures or humidity is a technique that can be used to destroy many stored-product pests. Heat treatment kills some pests outright; cold usually blocks their development. For adequate control, it may be necessary to subject products to a prescribed period of high temperatures followed by cold, after which they should be kept stored at a constant, lowered temperature. In general, a temperature of 60° F prevents insect feeding; 40° F kills insects over a period of time. Some products can be kept frozen to protect them from insect damage.

Desiccants

Dusts, such as silica gel or diatomaceous earth, can be combined with certain stored grains to provide protection against insect damage. These dusts kill target insects through desiccation. Dusts are removed from grain and other stored food items before processing by a cleaning operation that also removes other debris. Because sorptive dusts are inert, no potentially harmful residues are left on the food if traces of the desiccant remain.

Insecticides

Insecticide use varies according to the type of pest and situation where the infestation occurs. Because food products are involved, residues must never exceed legal tolerances. Apply only those insecticides registered for stored food products and use them in strict accordance with label instructions. Insect resistance to insecticides is an increasing problem, so avoid overusing insecticides and always employ other methods of control along with them. Make insecticide applications at times when insects are most susceptible.

The safest type of insecticides for use on food items are the microbials such as *Bacillus thuringiensis*. Those organisms produce toxins that are fatal to certain species of insects but have no known effect on people. Use only microbial insecticides that are labeled for control of stored-product pests and that can be applied directly to the product being treated. Thorough coverage is necessary to ensure that target insects consume some of the microbial organisms.

Insect growth regulators (IGRs) have a low toxicity to humans as compared with organophosphate, carbamate, or chlorinated hydrocarbon insecticides. IGRs are chemicals that alter an insect's ability to develop normally or pass through developmental stages at the proper time. For instance, some IGRs prevent larvae from becoming adults, and others force them to pass into the adult stage before they are mature enough to reproduce.

Because of the low toxicity of IGRs, they are usually safe to spray directly onto raw products (check the label before application). Use an IGR where fumigation is not possible or desirable. An IGR is only effective if it contacts the targeted insect pest, therefore thorough coverage is necessary. Apply a spray of a labeled IGR to grains, nuts, or other foodstuffs during the filling of storage bins. Use enough spray to thoroughly protect all of the stored product. Spray when insects are at the correct stage of development as described on the IGR label instructions. Occasionally the application of an IGR extends the larval period, and therefore larvae of pest insects may feed more before they are destroyed.

Fumigants are used to control stored-product insects in bulk containers, truck trailers and railroad cars, warehouses, and large storage areas. For complete information on fumigation, refer to the companion publication to this volume, *Fumigation Practices*. Fumigants are effective because they penetrate into areas where pests occur or might become problems. To be effective, fumigation must take place in a well-sealed area so the fumigant concentration can build up to high enough levels; other conditions must also be met and specific problems overcome before fumigation takes place (Table 8-1).

TABLE 8-1

Items in the Environment that May Affect a Fumigation. Some fumigants may react with items listed on this table. Check fumigant label for precautions.

PROBLEM	EXAMPLE	EFFECT
SORPTION	water/high-moisture soil/sand concrete charcoal mortar cinder blocks bricks	Absorbs or adsorbs fumigant molecules and reduces concentration in the environment. Requires using more fumigant and longer fumigation periods. May desorb, or give off fumigant molecules, at toxic levels after area has been aerated.
REACTION	open flames glowing electrical coils, stove heating elements	Fumigant molecules undergo chemical reaction to produce another compound that is corrosive or harmful to items in the fumigated area.
DESTRUCTION	seeds and bulbs nursery stock pets livestock	Some fumigants are toxic to living plants or animals and will destroy them.
DAMAGE	rubber furs horsehair leather woolens rayon vinyl paper cellophane photographic chemicals polypropylene foam or felt carpet padding others	Some fumigants react with sulfur or sulfides in items in the fumigated area to produce obnoxious odors. Some damage items in the fumigated area. Check fumigant label to be certain it can be used with items found in the area.
TOXIC RESIDUES	fruits vegetables cereals salt dried foods	To avoid toxic residues, remove food items from fumigated area. If food items are being expressly fumigated, use only labeled amounts and do not increase fumigation time. Follow label instructions carefully.

Small quantities of cereals and similar products can be fumigated in containers such as plastic pails or glass jars using dry ice (frozen carbon dioxide); however, if containers are tightly closed immediately after treatment, a vacuum will form that may cause them to implode. Tighten down the lid after the container warms to room temperature.

Short-term residual insecticides, such as pyrethrins or pyrethroids, can be used for rapid knockdown of some types of stored-product insects. Apply these materials in cracks and crevices and on surfaces that stored products contact. These materials can be applied to bulk containers before adding foodstuffs, for example. They are also used in cupboards and on shelves and areas close to where products are stored, but usually require frequent reapplication if infestations are high.

Residual insecticides, including some persistent pyrethroids, should be selectively used for control of stored-product insects. Residuals are generally applied to surfaces of empty containers to prevent pest infestation, but rarely applied directly to foodstuffs. Residual insecticides should be used as a supplement to sanitation measures. They are convenient ways to control stored-product pests in inaccessible areas.

There are severe restrictions on pesticide residues on food in food-handling establishments, so be sure residual insecticides are used only according to label instructions and in compliance with federal, state, and local regulations.

MITES

Mites occasionally infest stored food products. They are known to feed on cheese, flour, grains, dried fruits, dried meats, cereal foods, dog and cat food, and animal feeds. Grains often must first be damaged by insect or fungus pests before certain mite species can invade. There are over 112 species of mites commonly associated with stored foods. Because mites are extremely small, their presence goes unnoticed, but the damage they can cause is sometimes very serious. Infested items become contaminated with living and dead mites, cast skins, and fecal materials.

Feeding by some mite species alters the nutritional quality of grains and other food items; mites often attack the germ of grains. Flour from mite-damaged grain may become sour and have poor color, and bread made from it does not rise properly.

Some mites are fungus feeders. They invade commodities that have become moldy, bringing spores of certain fungi, and feed on the fungi once they become established. Even after the mites are controlled, the fungi persist and continue to cause damage.

Management Guidelines for Stored-Product Mites. The most difficult part of managing stored-product mites is detecting a mite infestation. Large populations can develop before their presence is discovered, and by the time an infestation can be seen considerable damage may already have taken place. The stored food product may take on an odor variously described as minty, sweetish, or musty when it is infested with mites. This odor may be the first indication that mites are present.

Use a microscope or hand lens to inspect stored products for the presence of mites. Under magnification, these small, colorless or cream-colored mites can be seen moving about. Take several samples throughout the stored product and examine each one carefully. Check for moldy areas and for mites associated with the fungus.

Avoid attracting mites by using sanitation practices to eliminate residues around the storage facility. Clean storage containers before their use to remove debris and mites and mite eggs. Inspect materials before they are put into the storage facility to be sure they are pest-free. Maintain proper storage conditions, including controlling the moisture and air circulation to prevent the growth of fungi. Keeping the stored product at or below a moisture content of 12% also retards development of many species of mites.

Desiccants, fumigants, and some types of residual sprays effectively control mites as long as the commodity has been uniformly treated. Usually insecticidal treatment of the commodity or storage container for control of stored-product insects also destroys mite infestations. Periodic retreatment may be necessary after an infestation has been controlled because mite eggs may not have been destroyed. Check the label of the pesticide for permitted uses and follow label instructions carefully.

9 Wood-Destroying Pests

Wood-destroying pests include termites, carpenter bees, ants, wood-boring beetles, certain fungi, and marine borers. These are important economic pests because they damage or destroy structural wood —the wood used in homes, apartments, offices, and warehouses—as well as decorative wood, piers and pilings, posts, furniture, and sometimes plastics. At least 1% of the housing units in the United States require treatment each year for control of termites; building owners spend nearly $2 billion yearly for termite control or repair of damage. Damage from wood rot fungi, wood-boring beetles, carpenter bees, and ants further increases these losses.

TERMITES

Termites belong to the insect order Isoptera. These are primitive insects believed to be closely related to cockroaches, probably sharing a common ancestor. Cockroaches, however, are in a different insect order. Like cockroaches and other primitive insects, termites do not undergo complete metamorphosis. Young, known as nymphs, are fed and groomed during part of their development by other members of the colony. Termites have the most highly developed insect social structure, living in large colonies in the soil or in chambers carved in dead or, sometimes, living wood. Colonies are composed of castes, specialized forms of individuals that include soldiers, reproductives, and, in some species, workers. Unlike social bees and wasps, each caste is made up of members of both sexes.

Soldiers have greatly enlarged heads and mandibles which they use to defend their colonies. Workers, the most numerous caste in colonies of many termite species, are responsible for constructing living chambers and tunnels and foraging for food. They also groom and feed one another and other colony members. Other termite species, such as the drywood termites, are believed to be more primitive; they do not have a worker caste, so these functions are carried out by immature soldiers. Workers of the more advanced species probably evolved from this soldier caste. Reproductives are long-lived queens and kings which are winged during their early adult life but lose their wings after dispersing from their original colony.

Communication among colony members takes place by means of pheromones. Individuals produce pheromones to mark trails, which are followed by other colony members. Hormones excreted into feces regulate the development of individuals that consume these materials while grooming, and are responsible for caste determination and suppression of reproductive functions.

Most species of termites have protozoa within their intestines that convert wood cellulose into sugar, allowing termites to feed on wood or paper. Newly hatched nymphs and freshly molted individuals do not have these protozoa but obtain them by feeding on excrement of other members of the colony.

Termites are long-lived and successful insects. Their success is probably because they exploit a large, almost unlimited supply of food and live hidden away in obscure places, protected from enemies and environmental

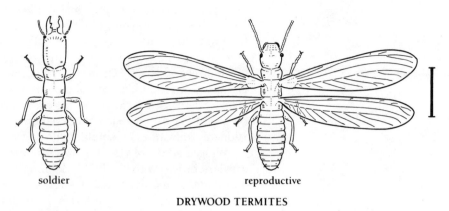

soldier   reproductive

DRYWOOD TERMITES

FIGURE 9-1.

Western drywood termite, Incisitermes minor.

extremes. Termite pests in California include the drywood termites, the damp-wood termites, and the western subterranean termite. These pests cause serious damage to wooden structures and posts and may also attack stored food, household furniture, and even plastics.

Drywood Termite
Kalotermes **spp.**

Western Drywood Termite
Incisitermes minor

Desert Drywood Termite
Marginitermes hubbardi

Drywood termites (Figure 9-1) are believed to be a more primitive species because they do not have a worker caste; nymphs destined to be soldiers assume this role. Drywood termites remain entirely above ground and do not connect their nests to the soil. They have low moisture requirements and are able to tolerate dry conditions for prolonged periods. Although colonies are small, they are long-lived and may exist for decades. An established colony contains several thousand individuals, compared to millions of individuals found in colonies of some other species. Colonies grow slowly and may have only 20 members after the first year. Drywood termites are the most typical termite in southern California, but also occur along most coastal regions, the central valley, and southern desert.

Drywood termites infest dry, undecayed wood. This includes structural lumber as well as dead limbs of native trees and shade and orchard trees, utility poles, posts, and lumber in storage. From these areas, winged reproductives seasonally migrate to nearby buildings and other structures. Mating and dispersal flights take place on sunny days during fall months. Usually, mated pairs enter buildings through attic vents, roof shingles, or openings around doors and windows. In hot, dry locations, they enter through foundation vents. An infestation begins when the mated pair finds suitable wood and constructs a small chamber, which they enter and seal. Soon afterward, the female begins egg laying. Newly hatched nymphs start enlarging the nest.

Although colonies are commonly established in roof structures and attics, the substructure of a building may be invaded by mated pairs entering through foundation vents. Occasionally they are found in wooden boxes, crates, or furniture; because colonies are small and hard to see, they can be easily transported to other locations.

A sign of drywood termite infestation is the appearance of piles of fecal pellets coming from openings in infested wood known as kickouts. Workers bore open kickouts and push accumulated debris out of the colony's tunnels and galleries. Fecal pellets have a distinctive appearance, helping to distinguish drywood termites from subterranean or dampwood termites. Pellets are elongate with rounded ends, about 0.8 mm long, and have six flattened or roundly depressed surfaces separated by six longitudinal ridges (see Figure 9-11 on page 174). Another sign of infestation is swarms of winged reproductives suddenly appearing in a building during warm fall days.

Winged adults of western drywood termites are about ½ inch long. They are dark brown with smoky black wings and have a reddish brown head and thorax; wing veins are black. These insects are noticeably larger than subterranean termites. Winged forms of the desert drywood termite are pale. Soldiers of this species have a clublike third antennal segment which is almost as long as all the succeeding segments combined, easily distinguishing them from other species.

Western Subterranean Termite
Reticulitermes hesperus

The western subterranean termite (Figure 9-2) is likely the most destructive of all termites found in California. This is due to (1) the large number of individuals in a colony, (2) their habit of nesting in the soil, (3) the difficulty in detecting and controlling them, and (4) occurrence of much of their damage in foundation and structural support wood. Colonies include reproductive, worker, and soldier castes. Reproductive winged forms of subterranean

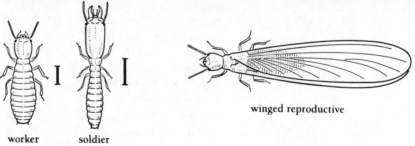

worker   soldier   winged reproductive

WESTERN SUBTERRANEAN TERMITE

FIGURE 9-2.

Western subterranean termite, Reticulitermes hesperus.

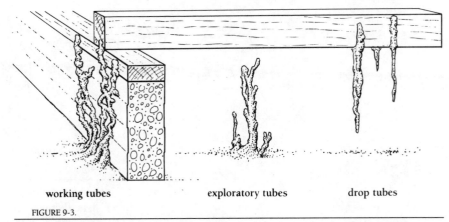

working tubes   exploratory tubes   drop tubes

FIGURE 9-3.

Subterranean termites construct four types of tubes or tunnels. Working tubes (left) are constructed from nests in the soil to wooden structures; they may travel up concrete or stone foundations. Swarming tubes are further extensions of the working tubes. Exploratory and migratory tubes (center) arise from the soil but do not connect to wood structures. Drop tubes (right) extend from wooden structures back to the soil.

termites are dark brown to brownish black, with brownish gray wings. They are about ³/₁₆ to ¼ inch long, excluding wings. Soldiers have white bodies with pale yellow heads. Their heads are long, narrow, and large, without eyes. Workers are slightly smaller than reproductives, wingless, and have a shorter head than soldiers; they are colored similar to soldiers.

Subterranean termites require moist environments. To satisfy this need, they usually nest in or near the soil and maintain some connection with the soil through tunnels in wood or through shelter tubes which they construct. Four types of shelter tubes are made by subterranean termites (Figure 9-3). *Working tubes* are constructed out of soil that the termites bind with liquid fecal material and an excreted gluelike substance; these tubes connect nests located in the soil to wooden substructures, providing safe travel for workers by protecting them from natural enemies. During dry periods, workers must periodically return to the soil to replenish their body moisture. *Exploratory* or *migratory tubes* also arise from the soil, but do not usually connect to wooden structures above; they are probably uncompleted working tubes, abandoned after workers were unable to locate a food source. *Suspended* or *drop tubes* arise from wooden substructures and travel down to the soil. They are lighter in color than working and migratory tubes because they are made of wood particles rather than soil; after establishing tunnels in wooden structures, drop tubes are constructed to give workers additional return routes to the nest or to humid soil. *Swarming tubes* are specialized shelter tubes used as nest exit routes by reproductives during swarming flights; they arise from the soil and may extend a considerable distance up on wooden structures.

Signs of infestation include swarming of winged forms shortly after late fall and early winter rains. Swarming sometimes extends into spring due to late spring rains. People sometimes confuse winged termites with winged ants, but closer examination easily reveals the differences (Figure 9-4): the hind wings of termites are nearly as long as the forewings, whereas winged ants have much shorter hind wings, and the termite abdomen is broadly joined to the thorax, whereas winged ants have a very thin waist joining the abdomen to the thorax.

Western subterranean termites feed along the soft parts of wood and usually do not cross harder grains. This species produces liquid fecal material

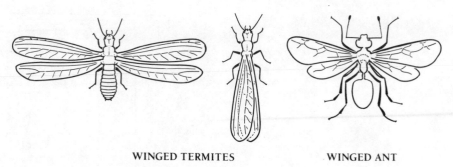

WINGED TERMITES   **WINGED ANT**

FIGURE 9-4.

Winged termites are distinguished from winged ants by their longer hind wings and the broad connection of their abdomen to their thorax.

rather than dry pellets, so tunnels are characteristically stained by the liquid feces. Darkening or blistering of wooden members is another indication of an infestation; wood in damaged areas is typically thin and easily punctured with a knife or screwdriver.

Pacific Dampwood Termite
Zootermopsis angusticollis

Nevada Dampwood Termite
Zootermopsis nevadensis

Like drywood species, dampwood termites are a primitive group. They do not have a social organization as highly structured as the subterranean termites. These species have only two distinct castes, the reproductives and the soldiers (Figure 9-5); nymphs perform the tasks of the worker caste.

Dampwood termites nest in wood buried in the ground, although contact with the ground is not necessary when infested wood is high in moisture. These termites sometimes attack decaying wood. Sound wood in both dead and living trees is attacked by dampwood termites; they usually gain entry through the roots. Because of their high moisture requirements, dampwood termites most often are found in cool, humid areas along the coast and are typical pests of beach houses. The Nevada dampwood termite, however, occurs in the higher, drier mountainous areas of the Sierra where it is an occasional pest of mountain cabins and other forest structures; it also occurs along the northern California coast.

All species of dampwood termites produce distinctive fecal pellets, which can be used for their identification (see Figure 9-11 on page 174). Dampwood termite pellets are rounded at both ends, elongate, but lacking clear longitudinal ridges; flattened sides are noticeable. Pellets are expelled through kickouts, although in damper areas they clump together and remain stored in unused tunnels. Winged reproductives typically swarm between July and October, but it is not unusual to see them at other times of the year.

Dampwood termites are the largest of the termites occurring in California. Winged reproductives are nearly 1 inch long and are dark brown with brown wings. Soldiers have a flattened brown or yellowish brown head with elongated black or dark brown mandibles. Nymphs are cream-colored with a characteristic spotted abdominal pattern caused by food in their intestines. Nevada dampwood termites are slightly smaller and darker; reproductives are about ¾ inch long.

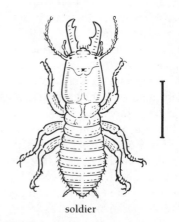

soldier

PACIFIC DAMPWOOD TERMITE

FIGURE 9-5.

Pacific dampwood termite,
Zootermopsis angusticollis.

Management Guidelines for Termites

Successful termite management requires many special skills, including a working knowledge of building construction. This will help you locate infestations and spot conditions that promote problems. You must also understand the biology and habits of each species of termite and the probability of their occurrence in your area. Termite identification and biology is important to know before beginning a control program because management techniques vary depending on the species causing an infestation. Also, more than one species of termite can infest a building at the same time, which influences your control approach. Subterranean and dampwood termites have similar ground-nesting habits, so control methods for these are usually similar. However, drywood termites nest above ground, therefore the approach for eliminating them is unique.

Inspection. Before beginning a control program, thoroughly inspect the building to obtain information about the infestation. Verify that there are termites, identify them, and locate the extent of their infestation and damage. Look for conditions within and around the building that promote termite attack, such as excessive moisture or wood in contact with the soil. Follow the procedures listed in Table 9-1 for inspecting a building for termite infestation. Using graph paper, prepare a floor plan or sketch of the building and indicate areas of damage, species found, and problems with the construction

TABLE 9-1

Termite Inspection Checklist.

Check to see that the building meets the following conditions. Variations from these conditions could subject the building to termite invasion or indicate the presence of termites.

FOUNDATION
General Inspection
1. Surface water drains away from building.
2. Rainwater from roof drains away from building.
3. Walkways, patios, porches, and slabs drain away from building.
4. Firewood and lumber is piled more than 6 inches away from building.
5. All wood around foundation of building that is in direct contact with the soil is pressure treated.
6. Wood siding is more than 6 inches above outside grade.
7. Paint or stucco around base of building is not blistered, peeling, or loose.
8. Untreated wood framing is more than 8 inches above outside grade.
9. No termite tubes are present on the foundation walls.
10. No evidence of decay or insect damage in wood siding, doors, door frames, sills, wooden steps, wooden columns, crawl space access doors, or wooden structures or utility accesses attached to building.
11. Form boards, grade stakes, wood debris, and paper products have been removed from beneath and around the building.

For Buildings with a Crawl Space
1. There is an access door and adequate cross ventilation in the crawl space (consult state and local building codes).
2. Tree stumps, untreated wood, wood debris, paper products, and plants have been removed from the crawl space.
3. No untreated wood blocks support ducts or pipes on the ground.
4. There is no evidence of standing water in the crawl space.
5. No clothes dryer vents or air conditioning condensate lines discharge into the crawl space.

6. No evidence of condensation on beams, joists, sills, and subfloor as indicated by fungal stains.

7. No evidence of leaks beneath kitchens, bathrooms, utility rooms, and other areas with plumbing.

8. Interior foundation walls are dry.

9. No termite tubes present on inside foundation walls, sills, plumbing, piers, fireplace foundations, or other areas.

10. No evidence of insect attack or decay in wood foundation walls or piers.

11. Wood beams are more than 12 inches above inside grade.

12. No evidence of insect attack or decay in wood beams.

13. Sills, joists, and subfloor more than 20 inches above inside grade.

14. No evidence of decay or insect damage on sills, joists, or subfloor.

For Buildings with Basements or Rooms Below Grade or Partially Below Grade

1. Untreated wood columns are not in contact with the slab nor penetrate the slab.

2. Wood in contact with the slab is pressure treated.

3. Basement or below-grade rooms do not show evidence of water leaks or excessive condensation on walls and floor.

4. No evidence of decay or insect damage on doors, door frames and sills, windows, window frames, baseboards, walls, or other wood structures.

ABOVE GROUND LEVEL

1. Windows are properly glazed.

2. Window and door frames properly caulked.

3. Flashing present above doors and windows, at intersections of different materials on exterior walls, at roof/wall intersections, at roof/chimney intersections, and at pipes and vents projecting through the roof.

4. Shingles extend at least ¾ inch beyond edge of roof at the eave and rake and form a continuous drip line.

5. Rain gutters are present if the roof overhang is less than 12 inches on a single story building or less than 24 inches on a taller building.

6. Gutters are cleaned of debris and do not leak.

7. Downspouts do not leak.

8. Attic is properly ventilated.

9. There is no evidence of decay or insect attack to any aboveground wood siding, doors or door frames, windows or window frames, or attached structures, including steps and porches.

INSIDE THE BUILDING AND ATTIC

1. No evidence of water leaks in bathrooms, kitchens, utility rooms, or other areas with plumbing.

2. No evidence of water stains or mildew or mold growth on ceilings or walls.

3. Floor is not sagged or buckled.

4. No gaps between floor and baseboards.

5. No evidence of insect attack or decay in windows or window sills, doors or door frames and sills, baseboards, flooring, walls, and other areas.

6. Access door to attic present.

IN ATTICS

1. No evidence of insect attack or decay on roof substructure, rafters, joists, wall top plates, or other wooden structures.

2. Vapor barrier of insulation is toward the living area of the building.

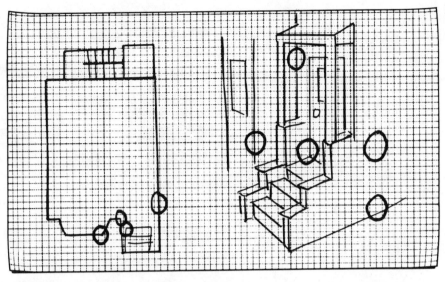

FIGURE 9-6.

Prepare a sketch of the building being inspected to indicate areas of termite damage and problems with the construction or physical condition of the building.

or physical condition of the building (Figure 9-6). Use sounding and probing procedures to help locate termite-damaged areas. *Sounding* is a technique of tapping on wood with a small hammer or other instrument to detect infestations; normal wood produces a sharp, clear knocking sound and tapping on infested wood produces a softer, dull response. Tapping may also cause soldiers of subterranean termites to knock their heads against sides of galleries to produce an audible warning sound. Walking through an infested area or opening a garage door may also elicit this response. *Probing* involves the use of a sharp instrument, like an ice pick, to detect tunnels beneath wood surfaces. Specially trained dogs can also be used for locating termites in buildings; their keen sense of smell often makes them successful in finding infestations.

Prevention. Use an integrated program to manage termites. Combine methods such as modifying habitats, excluding termites from the building by physical and chemical means, and using mechanical and chemical methods to destroy existing colonies. Building design may contribute to termite invasion (Table 9-2). Termite-resistant wood and other building materials may aid in reducing damage. Chemical treatment of structural wood used in foundations and other wood in contact with soil helps protect against termite damage in areas where building designs must be altered or concrete cannot be used. Follow Uniform Building Code guidelines for construction details.

Recent research has proved the effectiveness of foundation sand barriers for subterranean termite control (Figure 9-7). Sand with particle sizes in the range of 10 to 16 mesh is used to replace soil around the foundation of a building and sometimes in the crawl space. Subterranean termites are unable to construct their tunnels through the sand and therefore cannot invade wooden structures resting on the foundation.

Use screening over attic vents and seal other openings, such as knotholes and cracks, to discourage the entry of winged drywood termites. Remember that although screening of foundation vents or sealing other openings into the substructure helps block the entry of termites, these procedures may interfere with adequate ventilation and increase moisture problems.

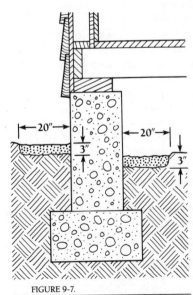

FIGURE 9-7.

Sand barriers have proved to be effective in preventing subterranean termite attack on building foundations.

TABLE 9-2

Areas of Faulty Construction that May Lead to Subterranean Termite Infestation.

PROBLEM	RESULT
FOUNDATION	
Low foundation walls or footings	Permit termites to reach wood of building.
Foundation cracks	Termites gain access to wood hidden from view.
Insufficient number or size of air vents	Keeps air under structure warm and moist, promoting ideal conditions for termites.
Wood debris left around or under structure	Supports termite colonies until they increase their population large enough to attack the structure.
Form boards left in place	Provides food for termite colonies and access into the building.
Improperly bonded brick veneer or stucco covering	Provides hidden entrance to wood portions of the building.
Faulty drainage or soil grade	Increases moisture around foundation, promoting ideal conditions for termites.
ATTACHED STRUCTURES	
Wooden porch steps in contact with the ground	Provides access for termites into the structure.
Porches, balconies, and landings not sloped away from building	Increases moisture at junction of structure and building, promoting rot and termites.
Gate posts, fence tie-ins, abutments, columns, and pilasters attached to structure and having contact with the soil	Provide access for termites to structural wood.
Planter boxes attached or built next to building without protective concrete barrier	Allow termites access to structural wood.
Trellises attached to structure and also in contact with soil	Allow termites access to structural wood.
OTHER EXTERIOR PROBLEMS	
Buckled or cracked siding or stucco	Admit moisture, increases potential for termite problems.
Warped or cracked sills or casings and window and door frames	Admit moisture, increases potential for termite problems.
Exposed wood junctions not properly sealed	Admit moisture, increases potential for termite problems.
Exposed beam ends	Admit moisture, increases potential for termite problems.
Leaking pipes or faucets	Keep wood and soil moist, promoting conditions for termites.
Badly fitted shingle roof	Admit moisture, increases potential for termite problems.
Improperly sealed eaves, roof overhangs, and fascia boards	Increase moisture and rot potential.
Poorly sloping rain gutters	Increase moisture problems leading to rot and termite damage.
ATTICS	
Inadequate ventilation	Increase temperature and moisture problems leading to rot and termite damage.

Keep all substructure wood at least 12 inches from the soil beneath the building. Identify and correct other structural deficiencies that attract or promote termite infestations. Keep attic and foundation areas well ventilated and dry. Inspect utility and service boxes attached to the building to see that they are sealed and do not provide shelter or a point of entry for termites. Reduce chances of infestation by removing or protecting any wood in contact with the soil. Inspect porches and other structural or foundation wood for signs of termites. Look for limbs and tree stumps, stored lumber, untreated fence posts, and buried scrap wood near the structure that may attract termites.

Buildings on concrete slabs or those that have no access to the interior parts of the foundation require injection of an insecticide through holes drilled under the slab or through the foundation.

Whenever possible, eliminate soil-exposed untreated structural wood. If concrete cannot be used, replace this wood with pressure-treated wood. If pressure-treated wood has to be cut or shaped, paint or dip cut surfaces with an appropriate preservative. Should it be impractical to replace soil-exposed untreated wood, paint or spray it with a suitable preservative. However, painted or sprayed surfaces are not as resistant to termite attack as pressure-treated wood. Pressure-treated wood offers greater protection against termites and other wood-destroying organisms because the pressure treatment enables deeper penetration of the preservative (Table 9-3).

When applying a preservative to wood surfaces, be sure to observe safety precautions and protective clothing requirements listed on the preservative label.

Apply silica aerogel to exposed wood surfaces in attics and other susceptible areas to prevent drywood termite damage. This desiccant provides long-

TABLE 9-3

Comparison of Depth of Penetration of Wood Preservatives Using Different Methods of Application. Actual depth depends on species of wood being treated and its moisture content.

APPLICATION METHOD	DEGREE OF PENETRATION
Brush on	Very shallow penetration—may not be uniform.
Spray on	Very shallow penetration—not always uniform coverage.
Dipping	Usually shallow penetration—penetration increases with length of time of dipping. Coverage is uniform.
Steeping or cold soaking	Moderate penetration. Requires long period of time. Uniform coverage.
Hot and cold bath treatment	Moderate to deep penetration. Uniform coverage.
Diffusion process	Moderate to deep penetration with uniform coverage. Process is very slow. Suitable only for water-saturated wood.
Vacuum process	Deep, uniform penetration, depending on length of time, amount of vacuum, and condition and species of wood.
Pressure treatment	Deep, uniform penetration, depending on length of time, amount of pressure, condition and species of wood, and other factors.

term protection as it does not break down under extreme temperatures. If used properly, it has low toxicity to people. However, it is a very fine dust and can cause serious respiratory distress if inhaled. Therefore, always wear a respirator capable of trapping extremely fine dust particles whenever applying silica aerogel. Do not apply it to areas where people or animals may come in contact with it.

Controlling Drywood Termites. Drywood termites can be controlled through physical removal of colonies, by freezing, through electrical treatment, or by fumigation or spot treatment of galleries with insecticides. The use of heat is another means of controlling drywood termites that shows promise and may reduce or eliminate the need for insecticides.

Drywood termite colonies are usually small, making it possible at times to control them by removing and replacing damaged wood. However, more than one colony may exist in a structure. Destroy damaged and infested wood promptly, preferably by burning if this is allowed. Otherwise, transport the material to a sanitary disposal site.

Drywood termites can sometimes be controlled by freezing the insects. Liquid nitrogen is injected into galleries to greatly lower temperatures.

Another method of control involves treatment of galleries with an electrical device called a termite electrocutor. This device uses high-voltage current to kill drywood termites where the infested wood is accessible. Electrical charges pass along termite galleries more readily than through undamaged wood because galleries offer a course of least resistance. The mode of action of the electrical current on drywood termites is unknown. It may take several days to several weeks after the treatment before individuals are killed. Special training from the manufacturer of this device is needed to learn the correct way to use it and to become aware of unique hazards and regulatory restrictions.

Insecticides used for control of drywood termites are applied as spot treatments to galleries or as total fumigation of the building. Only use pesticides labeled for termite control and for the type of site being treated. Follow mixing and application instructions carefully. Wear protective equipment required by the label.

If the infestation is not extensive and is isolated, spot-treat colonies with liquid or dust formulations. Inject insecticides into galleries through kickouts or holes drilled into wood surfaces. Use slow-acting compounds such as desiccants or dust formulations so individuals contacting the material will carry it back to the nest. Grooming activities spread desiccants or toxins to reproductives, soldiers, and nymphs even though the material does not reach them during application.

In buildings where drywood termite infestations are extensive, or if it is suspected that colonies are located in inaccessible areas, fumigation or heat treatment is generally required.

Heat treatment may control extensive infestations of drywood termites as effectively as fumigants. Heat treatment requires tarping the structure to prevent heat loss. A propane burner and large fan are used to heat and circulate air within the structures. Studies are still being carried out to determine the optimum temperature and amount of time required to achieve adequate control.

Refer to the supplemental publication *Fumigation Practices* for information on selecting and using fumigants. Buildings must be tarped or in some other way sealed to maintain high concentrations of fumigant so that it can penetrate termite galleries.

After fumigation or heat treatment, prevent reinfestation by sealing entryways into the building and by coating exposed surfaces in attics and other areas with desiccants (inert dusts or sorptive powders).

Controlling Subterranean and Dampwood Termites. Subterranean and dampwood termites cannot be properly controlled by fumigation, heat treatment, freezing, or termite electrocutor devices because the reproductives and nymphs are concentrated in soil nests below the structure, out of reach of these control methods. Whenever possible, destroy shelter tubes to interrupt access to wooden substructures and to open colonies to attack from natural enemies such as ants. If infestations are small, destroy accessible nests by repeated digging. Spot-treat nest areas with a liquid formulation of a long-lasting insecticide. If colonies are numerous or inaccessible, use soil drenches of a long-lasting liquid insecticide applied directly under the building or injected through the foundation or beneath concrete slabs.

When selecting insecticides, choose the least toxic material, yet one that will still be effective. Confine insecticide use to areas where termites are detected and to inaccessible areas. Chlorinated hydrocarbon insecticides, such as chlordane, have been used extensively for subterranean termite control because of their long persistence—30 years or more in the soil. Persistence and suspicions of health-related problems, however, have caused the use of chlordane to be banned. Pyrethroid and organophosphate insecticides, usually considered short-lived, are now being developed that will persist in soil for nearly 10 years. Foundations and structural wood can be protected by injecting insecticides into the soil beneath structures by horizontal or vertical drilling and rodding. Use insecticides only in accordance with label instructions and be certain that accurate dosages are applied.

Special hazards are involved with applying insecticides to the soil around and under buildings. Soil-applied insecticides must not leach through the soil profile to contaminate groundwater (mobility of insecticides in the soil is a chemical property that is also influenced by soil type, weather, and application techniques). Applications in the wrong place can cause insecticide contamination of heating ducts, radiant heat pipes, or plumbing used for water or sewage under the treated building (Figure 9-8). Check carefully for these types of hazards before injecting insecticides beneath or around a structure or into surrounding soil.

Experimental efforts have been made to control soil-dwelling termites through biological control. Argentine ants are a major enemy of termites and in some situations may be useful as control agents, especially of subterranean termites. Observations have shown that control is more effective if subterranean termite shelter tubes are opened and kept opened so ants have access to the colonies. Because ants are also building pests, their use as biological control agents requires methods to physically exclude them from building interiors.

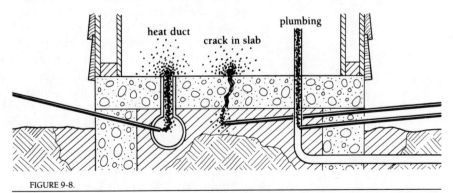

FIGURE 9-8.

Be sure to identify locations of heating ducts, plumbing, sewage lines, and electrical cables before drilling or injecting pesticides through walls or under slabs or foundations.

Certain species of nematodes have been used successfully in some instances to control subterranean termites, showing promise that these agents may prove to be suitable as biological control agents. Experimentally, parasitic nematodes in a water suspension have been applied in the same manner as liquid insecticides onto the soil near subterranean termite colonies. They are also injected into subterranean termite shelter tubes with an eyedropper or squeeze bottle. Upon finding termites, nematodes enter their bodies through natural openings. They carry bacteria with them which are released and invade termites' tissues, causing death. Nematode survival, and therefore success of this biological control technique, depends on dosage, soil types, specific temperature, and humidity requirements. If conditions are satisfactory, once nematodes find a termite they reproduce, then spread to other individuals. Because healthy termites eat dead or sick individuals, the nematode and bacterial infections are spread to other colony members. This control method has had only limited success, probably due in part to the narrow range of environmental conditions necessary for the nematode's survival, inadequate application techniques, failure to use sufficient numbers of nematodes, and the fact that the procedure is new.

WOOD-DESTROYING FUNGI

Wood decay is a serious problem in buildings, often closely associated with termite infestations and occasionally confused with termite damage. More damage is caused to structures by wood decay than by termites and wood-boring beetles combined. Moist conditions suitable for dampwood and subterranean termite infestations are also ideal for wood decay organisms. A condition called dry rot results from wood decay fungi. The name is inappropriate because moisture levels must be high at some time to support the growth of the fungus causing the rot.

Structural wood gets its strength from the compact arrangement of once-living cells that are tightly bound together by rigid cell walls. Wood decay fungi penetrate and destroy cell walls by dissolving them with enzymes. Cell walls and contents of cells are used by fungal organisms for food. Wood attacked by decay fungi, therefore, is soft and crumbly and breaks into many small square or rectangular pieces. Decayed wood usually has a dark brownish discoloration.

Fungi reproduce and spread by producing large quantities of spores, seedlike structures that develop into new fungal organisms. Spores are small and easily carried through the air, even on slight air currents. Spores can also be dispersed by free water such as rainfall or irrigation. Wood in contact with soil or rotting wood can easily pick up fungal spores. Spores are resistant to temperature extremes and can withstand long periods of time without moisture. When temperature and moisture conditions become adequate, however, spores develop into fungal organisms. Wood must have a regular moisture content of at least 20% before it can support fungal growth.

Only a few species of decay or rot fungi infest structural wood, although several other types of fungi are usually associated with them. Wood decay organisms include brown rot fungi, water-conducting or dry rot fungi, white rot fungi, and soft rot fungi. White rots can be distinguished by the bleached appearance imparted to the attacked wood; soft rots are found only in very wet situations and cause a gradual softening from the surface inward. The brown rot and water-conducting rot organisms are typically responsible for most structural damage; both of these impart a characteristic brown color to infested wood. Water-conducting organisms are different because they produce long, fibrous water-carrying structures called *rhizomorphs*. These rootlike structures enable fungal organisms to invade drier wood as far as 30 feet from a moisture source.

Mildews, blue-stain fungi, and slime molds are associated with wood decay fungi. Mildews cause discoloration of wood surfaces but do not themselves cause structural damage, although they can increase the tendency of wood to absorb moisture. Mildew stains can usually be removed by sanding or scrubbing. Blue-stain fungi produce a blue stain in sapwood that cannot be removed. Slime molds appear "egg white" on damp surfaces at certain times, while at other times they may resemble brightly colored mushrooms. Because they feed on spores and bacteria, the slime molds are important indicators of conditions suitable for the growth and development of wood decay fungi; however, they do not cause wood damage. Mildews and blue-stain fungi are also indicators of conditions that promote the growth of decay fungi.

Management Guidelines for Wood Decay Fungi. Control wood decay fungi by removing or eliminating the conditions that favor their development. Dry out moisture-laden wood to stop the damage. Locate moisture sources such as leaking plumbing, plugged rain gutters, poorly drained basements, or condensation in inadequately ventilated attics or crawl spaces and correct these faulty conditions (Table 9-4). Check suspected wood for damage by probing inward and upward with a sharp instrument such as an ice pick. Sound wood will break off in long splinters, while decayed wood comes out in short, brittle pieces. Remove damaged lumber and replace it with suitably treated wood or protect it from moisture by some other means. Cut away damaged wood up to several inches beyond fungal invasion. Use a moisture meter to determine potential for further problems. Moisture levels of about 20% in the remaining wood indicate conditions favorable for fungal infestations. Inspect surrounding lumber to be sure it is sound and not damaged.

Wood that cannot be kept free of moisture must be protected. Use paint or a water seal liquid to provide a waterproof barrier to windows, doors, siding, and eaves of buildings. Apply wood preservatives to wooden structural members of foundations where they contact moisture. Use posts that have been pressure treated with a preservative if they are to be buried directly in the soil; better yet, embed treated posts in cement. Damp areas must be well ventilated. The use of a moisture barrier under a building helps lower the humidity caused by moisture seeping up from the soil. Apply waterproofing materials to concrete and brick walls in basements and other areas where moisture passes through to wood.

Wood Preservation. Lumber used in structures can have natural resistance to insects and fungi, depending on the species of wood involved. The sapwood of all native wood species and the heartwood of most species are not resistant to decay organisms; even heartwood of resistant species is not totally immune to decay.

Resistance to decay is also accomplished by controlling the moisture content of the wood. Decay fungi cannot survive in wood that is water-saturated or has less than a 20% moisture content.

Sometimes lumber is protected from insects by covering it with mechanical barriers, such as metal sheathing, bricks, or concrete, after it is put into use (Figure 9-9). This technique offers some protection, but usually can-

FIGURE 9-9.

Masonry may be used to protect wood from moisture and insect damage.

TABLE 9-4

Typical Areas in Structures Where Wood Decay May Be Present.

LOCATION	MOISTURE SOURCES OR DEFECTIVE CONDITIONS
Bathrooms	Leaking plumbing.
	Condensation on pipes.
	Leaking shower pans.
	Leaking drain plumbing.
	Condensation of water vapor from use of hot water in tubs, showers, sinks.
	Defective toilet seal.
	Water frequently spilled on floor or splashed on walls.
	Defective seal between tub or shower and floor covering.
	Defective seal between sink and counter top.
	Broken or cracked wall or floor tiles, shower tiles, sink counters, or tub surround.
	Using too much water for cleaning floors or surfaces of counters.
Kitchens	Leaking plumbing.
	Leaking drains.
	Water frequently spilled on floor or splashed on walls.
	Condensation of water vapor from use of hot water or cooking.
	Defective seal between sink and counter top.
	Using too much water for cleaning floors or surfaces of counters.
	Water leaking from refrigerators, freezers, or ice makers.
Utility rooms	Leaking plumbing.
	Leaking drains.
	Water frequently spilled on floor or splashed on walls.
	Condensation of water vapor from use of hot water for laundry.
	Improperly ventilated clothes dryer.
	Using too much water for cleaning floors or surfaces of counters.
	Improper storage of wet clothing or other fabrics.
	Worn seals on washing machine.
	Leaks or condensation on water heaters.
Heating/cooling units	Moisture condensing on ducts or dripping from coils.
	Leaks in hot water or steam heating units.
General indoor areas	Water vapor condensing on windows.
	Leaking roofs.
	Improperly sealed exterior walls.
	Leaking indoor planters or pots.
	Improperly ventilated attic or crawlspace.
Outdoor areas	Blocked or leaking rain gutters or downspouts.
	Improper drainage of water away from building.
	Unprotected exterior wood or exposed cut ends.
	Irrigation system causing frequent wetting of building or area beneath building.
	Dense shrubbery blocking vents and air circulation next to building.
	Inadequate roof overhang or lack of rain gutters.
	Leaking planters attached to building.

not prevent damage from the fungi associated with moisture and may even increase decay problems by trapping moisture.

Chemical treatment of structural wood offers long-term protection from insects, fungi, and marine borers. Chemical treatment is the only successful and economic method of protecting wood subject to regular moisture contact. See the supplemental publication entitled *Wood Preservation Practices* for additional information on wood preservatives.

MISCELLANEOUS WOOD-DESTROYING PESTS

Carpenter ants, carpenter bees, and wood-boring beetles occasionally cause damage to structures and require control.

Carpenter Ants
Camponotus spp.

Several species of carpenter ants (Figure 9-10) are capable of damaging wood in buildings and other structures. These cause problems mainly in rural mountainous and forested areas, although they may also invade buildings in urban locations. Carpenter ants are easily distinguished from other species of ants by their large size, up to ½ inch long. Most species are dark, often black. Carpenter ants cannot sting, but if handled can inflict a painful bite with their powerful jaws. Some emit a noxious excretion of formic acid when disturbed.

Carpenter ants feed on dead and living insects, aphid and scale honeydew, and juices of ripe fruit. Carpenter ants enter buildings in search of food and may construct nests containing several thousand individuals somewhere within the building. Nests constructed indoors may be satellite colonies of a larger nest located outside near the building, usually in trees.

Although ants do not eat wood, they bore into wood to make their nests, which consist of extensive networks of galleries usually begun in areas soft from decay. Indoor carpenter ant nests are bored into wood parts of the building, sometimes causing serious structural damage. They also nest in wall voids, hollow doors, cracks and crevices, furniture, and termite galleries. Infestations can occur in new buildings when land clearing in the area disturbs existing native colonies. In the wild, carpenter ants nest in the soil and beneath rocks and bore into living and dead trees and tree stumps.

Management Guidelines for Carpenter Ants. Exclude carpenter ants from buildings by caulking cracks and blocking other entrances whenever possible. Trim branches and limbs of trees and shrubs that touch the building to keep ants from gaining access from these routes. Eliminate food sources inside the building or prevent access to suitable food by keeping it in antproof containers. Use a nonorganic mulch (such as gravel or stones) around the perimeter of the building to discourage nest building. Locate and destroy colonies in tree stumps and other nearby places. Eliminate damp conditions that promote wood decay. Replace decayed or damaged wood and correct problems that caused the decay, such as clogged rain gutters. Increase ventilation to damp areas beneath the building and in attics. Store firewood up off the ground and several feet away from buildings to discourage carpenter ant colonies.

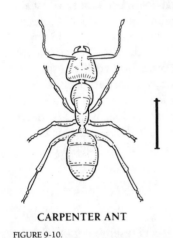

CARPENTER ANT

FIGURE 9-10.

Carpenter ant,
Camponotus *spp.*

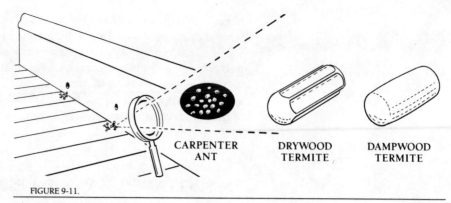

FIGURE 9-11.

The sawdust produced by carpenter ants is distinctly different from frass produced by termites.

Search for nesting sites in the building. Nests may be located by observing ant activity and following their trails, especially during the night. Try to find the gallery openings, which are usually small oval holes. Look for sawdust accumulations associated with these openings. Carpenter ant sawdust is considerably different from the pelletized frass left by drywood termites (Figure 9-11). Once colony openings are located, apply desiccant dusts or insecticide formulations (liquids or dusts) through these and other holes drilled into galleries. Use slow-acting formulations to ensure that they are carried back to reproductives and larvae deep inside the nest. If these methods fail because there is an extremely heavy infestation or nests are not accessible, it may be necessary to fumigate the entire building. Replace severely damaged structural wood and physically remove nests; use a vacuum to collect individual ants.

Carpenter Bees
Xylocopa spp.

Most carpenter bees are large and robust insects resembling bumble bees (Figure 9-12). They are usually about 1 inch long and colored metallic blue black with green or purplish reflections. They differ from bumble bees by having a shiny, hairless abdomen. Males of some species are lighter, ranging into golden or buff hues.

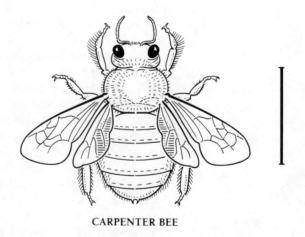

CARPENTER BEE

FIGURE 9-12.

Carpenter bees, Xylocopa *and* Ceratina *species.*

The presence of carpenter bees around buildings and wooden structures can be annoying or even frightening, although males cannot sting and females rarely attack. However, these bees cause damage to wooden structures by boring into timbers and siding to prepare nests. Sound, undecayed wood without paint or bark is usually selected for nests; carpenter bees frequently attack southern yellow pine, white pine, California redwood, cedar, Douglas fir, and cypress. They avoid most harder woods.

Nests usually consist of tunnels ½ inch in diameter and 6 to 10 inches deep, partitioned into several chambers, each containing an egg and a supply of food. Carpenter bees may use old tunnels for their nests, and sometimes enlarge these; several bees may use a common entry hole connecting to different tunnels. Over a period of time, tunnels may extend as far as 10 feet into wood timbers. Carpenter bee nests weaken structural wood and leave unsightly holes and stains on building surfaces. Tunnels are vacated after the brood's larval and pupal stages and remain empty throughout the rest of the year.

Management Guidelines for Carpenter Bees. Prevention is the main approach to managing carpenter bees. If possible, susceptible exterior parts of a building should be constructed out of harder wood varieties not normally attacked by the bees for nests. On all buildings, fill depressions and cracks in wood surfaces so they are less attractive. Paint or varnish exposed surfaces regularly to reduce weathering. Fill unoccupied holes with steel wool and caulk to prevent their reuse. Wait until after bees have emerged before filling the tunnels. Once filled, paint or varnish the repaired surfaces. Protect rough areas, such as ends of timbers, with wire screening or metal flashing.

Carpenter bees are generally considered beneficial insects because they help pollinate various crop and noncrop plants. Under most conditions they can be successfully controlled using the preventive measures described above. If infestation is high or risk of damage is great, insecticides may be used to augment other methods of control. To do this, treat active nests (those containing eggs, larvae, or pupae) with desiccant dusts or liquid or dust formulations of insecticides. After the brood is killed, repair holes with steel wool and wood filler, then repaint or varnish the repaired surfaces.

Wood-Boring Beetles

Wood-boring beetle larvae feed on wood and wood products; adults emerge from larval feeding chambers through round, oblong, or D-shaped exit holes. Adults of some species also bore holes into plaster, plastic, and soft metals. Many species of wood-boring beetles, especially those in the family Buprestidae (flatheaded or metallic wood borers) or the family Cerambycidae (which include the long-horned beetles and roundheaded wood borers), feed on living trees but do not reinfest lumber or wood products. Three families of beetles have species of wood borers that invade and damage structural and decorative wood and furniture. These families are the Lyctidae, Anobiidae, and Bostrichidae.

Powderpost Beetles
Family *Lyctidae*

Beetles in the family Lyctidae are known as powderpost beetles (Figure 9-13). They are so named because larvae leave a fine, dustlike powdered frass in their

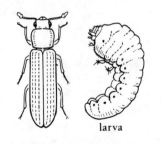

larva

POWDERPOST BEETLE

FIGURE 9-13.

Powderpost beetle, Lyctus *spp.*

galleries which occasionally falls out of exit holes into small piles on floors or other surfaces. This powdered frass distinguishes powderpost beetles from the other typical wood-boring beetles. Lyctids attack hardwood species apparently because these woods have pores into which they can oviposit; softwoods do not have such pores. Also, the starch content in softwoods is nutritionally low for these beetles.

Adult lyctids range from $1/12$ to $1/3$ inch in length, depending on species, and are usually brownish to reddish. They most often attack oak, ash, hickory, mahogany, and walnut. Infestations may occur if beetles or larvae are brought into a building in furniture or firewood. Sometimes the only sign of infestation may be the tiny, round exit holes made by emerging adult beetles. Once they emerge, the winged adult beetles spread to other wood surfaces where they deposit eggs onto unfinished surfaces or in cracks or other openings. They have a life cycle ranging from 3 months to over 1 year, depending on temperature, humidity, and the nutritional quality of the wood.

Deathwatch Beetles
Family *Anobiidae*

Wood-boring beetles in the family Anobiidae are known as deathwatch beetles (Figure 9-14). Deathwatch beetles are closely related to the drugstore beetles described in Chapter 8. Adults communicate with each other and probably locate mates by tapping their heads against wood, usually at night. (Deathwatch beetles supposedly acquired this name from people who have heard the tapping while sitting up with a sick or dying person during the night.) Adults are $1/6$ to $1/4$ inch in length and reddish to dark brown. Adults lay eggs in crevices or small openings or pores in unfinished wood. Two years may be required to complete each generation.

Deathwatch beetles are found primarily in soft woods, including girders, beams, foundation timbers, and some types of furniture. Some species attack books. This beetle is typically found in old wood and may be associated with wood that is partially decayed. Larvae of deathwatch beetles fill their galleries with small pellets of frass—smaller than the pellets produced by drywood termites—which distinguish them from other wood borers. None of the other boring beetles produce pelletized frass.

False Powderpost Beetles
Family *Bostrichidae*

Wood-boring beetles in the family Bostrichidae (Figure 9-15) are sometimes known as false powderpost beetles. Larvae tightly pack their galleries with frass that has the consistency of coarse powder; this coarse texture distinguishes them from true powderpost beetles as well as the deathwatch beetles. Adults are about $1/4$ inch long, mostly dark brown or black, sometimes with reddish mouthparts, legs, and antennae. Some species are large, adults reaching $1\frac{1}{2}$ to 2 inches in length. Adult beetles have a humpback appearance, so their head is not visible when viewed from above. This characteristic is also seen in some other wood-boring beetles. Females bore a tunnel, or egg gallery, into wood or other materials, then deposit their eggs in pores or cracks within the tunnel. Adults of some species bore through soft metal, such as lead and silver, as well as plaster and other nonwood materials, searching for sites to deposit eggs or for protection from weather extremes. This

**CALIFORNIA
DEATHWATCH BEETLE**

FIGURE 9-14.

Deathwatch beetle, family Anobiidae.

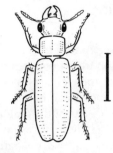
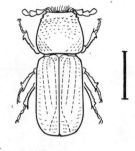

STOUT'S BOSTRICHID   LEADCABLE BORER

FALSE POWDERPOST BEETLES

FIGURE 9-15.

False powderpost beetles, family Bostrichidae. *Leadcable borer,* Scobicia declivis *(right).*

gives rise to the common name "leadcable borer," given to one species because of its habit of boring into the metal covering of suspended telephone wires. In buildings, the false powderpost beetles infest floors, furniture, hardwood paneling, and other wood materials.

Management Guidelines for Wood-Boring Beetles. Wood-boring beetles are difficult to control once infestation has begun. Prevention is the best management method, and protective measures should be taken at every stage of lumber processing and handling including lumber mills, plywood mills, lumber yards, furniture manufacturing factories, and building construction firms. Sanitation is the most important aspect of prevention. Remove and destroy dead tree limbs around buildings or near any area where wood products are stored. Destroy scrap lumber and other wood products before they become infested. Kiln drying of lumber destroys beetle infestations, although it does not prevent reinfestation. Materials used for construction of buildings and wood furniture should be thoroughly inspected before use to be certain that they do not contain wood-boring beetles. Protect wood from infestation by painting or varnishing to seal pores, cracks, and holes where eggs could be laid.

To keep from accidentally introducing wood borers, inspect furniture and other objects such as firewood before bringing them into buildings. Fumigate objects that show signs of beetle infestation. Also, inspect the building for signs of wood borer damage. Look for exit holes where adult beetles have emerged. Once galleries have been located, tap out frass to aid in identifying the pest.

Remove and replace infested structural wood whenever possible to eliminate beetles. Destroy infested wood by burning or take it to a landfill area. Use liquid insecticides where removal is not possible; apply the insecticide only to infested areas, but be sure these are thoroughly soaked. Remove and fumigate infested furniture. Extensive infestations within a building, or where inaccessible structural parts of a building are involved, may require whole-building fumigation. Whenever applying dusts, liquid insecticides, or fumigants, be sure to follow label instructions carefully. Wear the required protective clothing and respiratory equipment.

Marine Borers

Wood exposed to coastal or brackish waters is subject to damage or destruction from two types of marine borers: molluscan borers, distantly related to oysters and clams, and crustacean borers, relatives of lobsters and crabs.

Some molluscan borers actually resemble clams; however, these occur only along the coasts of the Gulf of Mexico and Hawaii. Molluscan borers found along the Pacific coast are wormlike in appearance and are commonly called shipworms. Shipworms hatch from eggs into free-swimming larvae that search for suitable wood, into which they make a small entrance hole. Once inside, shipworms enlarge the hole into a chamber where they remain throughout their life. They feed on wood from within the chamber and on minute organic particles, including plankton, found in seawater. Large infestations of these molluscan borers severely weaken or destroy pilings, wooden boats, docks, and other wooden objects.

Crustacean borers are distinctly different from molluscan borers in appearance and habits as well as the damage they cause to wood. They do not become imprisoned inside the wood they invade. They make narrow galleries that seldom extend very far below the surface. Under conditions of heavy infestation, the outer shell of the attacked wood becomes thoroughly honeycombed. This layer is gradually eroded away due to wave action and the battering of floating debris, exposing new wood to attack. The location of attack is usually limited to an area on pilings between half-tide and low-tide levels. Prolonged attack and erosion gives pilings a characteristic hourglass shape. Eventually, enough wood is removed to make the wooden structure useless.

Protection against marine borers is achieved by using properly treated wood. Creosote or creosote compounds have proved to be highly effective materials for protecting pilings, docks, and other wooden structures in coastal or brackish waters. These materials prolong the useful life of wooden structures to 15 to 30 years. Without treatment, wooden structures may be destroyed within 6 months to a few years.

Special paints are used to protect wooden boats from borer damage and to repel other crustaceans.

10 Vertebrate Pests

The most common vertebrate pests found in or around buildings are rats, mice, bats, and birds. Occasionally other vertebrates such as snakes, lizards, skunks, and squirrels enter buildings, although these are usually nuisances rather than destructive pests. Rats and mice, however, are troublesome because they are well adapted to people's habitats and they destroy or contaminate food and fabrics, cause structural damage, and are primary or intermediate carriers of transmissible disease organisms.

Bats and birds do not live intimately with people, but many species use buildings for suitable roosts or nests. They produce smelly and unsightly urine and droppings, create noise, build messy nests, and may have several insect and mite pests living on their bodies or in their nests. There is also concern that bats can be infected with rabies. Birds may be capable of harboring several different diseases; the disease most often associated with birds in buildings is histoplasmosis.

Managing vertebrate pests requires special skills because these animals are larger and more intelligent than invertebrates. Vertebrates may learn to recognize and avoid some control attempts, therefore an integrated and persistent approach is needed. It is necessary to understand life habits and food preferences, for instance, in order to monitor and manage population levels effectively and economically; otherwise you may get poor control or even increase the problem.

RATS

The Norway rat and roof rat are the principal rat species that may be found in buildings. The two species can be distinguished by certain physical characteristics including relative size of body and tail, shape of nose, and size of ears and eyes (Figure 10-1).

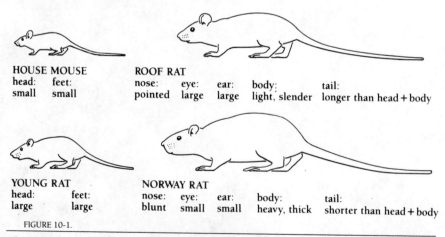

HOUSE MOUSE
head: small   feet: small

ROOF RAT
nose: pointed   eye: large   ear: large   body: light, slender   tail: longer than head + body

YOUNG RAT
head: large   feet: large

NORWAY RAT
nose: blunt   eye: small   ear: small   body: heavy, thick   tail: shorter than head + body

FIGURE 10-1.

The Norway rat and roof rat can be identified by certain physical characteristics including relative body size, shape of nose, size of ears and eyes, and length of tail.

Rats are destructive building pests. They eat and contaminate food products such as fruit, vegetables, grains, and meat; packaged food or beverages in cardboard, paper, foil, plastic, or cloth containers; and pet food. They also destroy textiles, upholstery, paper, books, and insulation by using these materials for making nests. Rats create holes in walls and around doors or windows and gnaw on electrical wiring, water pipes, and gas lines. Their gnawing can cause fires and other forms of severe damage. They leave droppings, urine, and hairs wherever they wander.

Several human diseases are associated with rat infestations. Salmonellosis is a serious intestinal disorder that can be transmitted to people who ingest food contaminated by salmonella bacteria in rat urine or feces. Murine typhus, leptospirosis, listeriosis, and trichinosis are other rat-transmitted diseases. The tropical rat mite, an external parasite of rats, causes severe itching and skin irritation in people. Rats have occasionally been known to bite sleeping people; bites can result in an infection known as rat bite fever. Plague bacteria can be transmitted from rats to people through the bite of rat fleas.

Rats gain entry into buildings several ways; they require openings of ½ inch diameter or larger. They may get in through broken windows, poorly screened attic and foundation vents, openings through walls used for passage of electric, gas, water, and sewage services, and through other openings or cracks in foundations, walls, or roofs. They can also chew holes through wood window or door frames. Dilapidated or poorly maintained buildings usually have many places for rats to enter. Poor building design or construction also contributes to infestations. Buildings serve as sources of food, water, shelter, and protection from natural enemies. For roof rats, shrubs and trees growing near buildings, including fruit and nut species, furnish attractive nest areas and abundant food; thick plantings of vines such as ivy make ideal locations for nests. Trash and garbage stored near buildings supply food and nesting sites, encouraging rat populations. Poor housekeeping within buildings also contributes to conditions that favor rat infestations.

Rats sometimes live in sewer systems and use these as travel routes for access into buildings. They get into sewers through poorly covered drains, broken lateral lines, and roof vents; rats have been known to come up through toilet bowls after coming up through sewer pipes or climbing down the inside of a vent pipe. The ample food, water, and shelter provided by some sewers allows rat populations to build up to large numbers; individuals may then invade buildings for food or in search of new shelters.

Rats have poor vision but highly developed senses of smell, taste, hearing, and touch. They use their senses to locate food and avoid danger. Rats forage for food in buildings mostly during the night. This behavior usually avoids encounters with people. Rats are agile and are able to run quickly and climb and swim well. They squeeze through small openings to get to food or escape from danger. These rodents are extremely wary of new items or situations and will sometimes take several days to adjust to changes in their environment before they investigate new food or nest items.

Norway Rat
Rattus norvegicus

The Norway rat, also known as the brown rat, house rat, wharf rat, and sewer rat, is the largest of the two rat species commonly found in buildings (Figure 10-2). Adults weigh between ¾ and 1¼ pounds; their average length ranges between 7½ and 10 inches, excluding their tail. Tail length is less than head and body length, and ranges between 6 and 8½ inches. Norway rats have coarse brown fur with scattered black hairs. The underside of the body

FIGURE 10-2.

Norway rat, Rattus norvegicus.

is usually gray, but may also be shades of yellowish white. The almost hairless tail is colored dark brown above, lighter below. This species has small, closely set ears, a blunt muzzle, and small eyes.

Norway rats become sexually mature at 3 to 5 months of age. They live for an average of 9 to 12 months in the wild, although their life span in captivity may be much longer. Females produce four to seven litters of 8 to 12 young each.

Outdoors, Norway rats usually nest in the ground. They construct their burrows under cement slabs, in lumber piles and garbage and rubbish heaps, along stream banks, or in other suitable locations. Nests are often no more than 6 to 8 inches below the surface and may be connected to *bolt holes,* separate exits used for escape when the main tunnel is blocked or endangered. From outdoor nests, Norway rats forage for food, entering buildings at night but returning back outside before dawn. During cold or rainy weather, Norway rats tend to move into buildings in search of shelter, warmth, and food. Indoors, they nest in secluded areas such as wall voids, behind appliances, beneath floors, and in drawers and closets where they will not be disturbed. When food and shelter requirements are adequate, these rats remain in a building throughout their life; under other conditions, many migrate outdoors and even away from the building as weather improves.

Norway rats generally feed on any type of food, but prefer greasy meat and animal products as well as fruits, grains, and vegetables. When starved, they eat other items such as soiled or stained clothing, snails, cockroaches and other insects, and animal feces. Water can often be obtained in buildings from leaking pipes or faucets, condensation, pet dishes, and sinks and toilet bowls. Norway rats have also gnawed holes in plastic and metal pipes to obtain water.

Roof Rat
Rattus rattus

Roof rats (Figure 10-3) are smaller than Norway rats, but have a tail longer than the combined length of the head and body. In adults, head and body length ranges from 6 to 8½ inches and tail length is from 7 to 10 inches. Body weight ranges between ½ and ⅔ pound. Tail color is uniform on both sides rather than being lighter on the underside. There are several color variations in roof rats, ranging from mottled grayish white with white underbelly to solid black with a gray underbelly. Roof rats have a pointed muzzle and large prominent ears; eyes are larger and more pronounced than those of

FIGURE 10-3.

Roof rat, Rattus rattus.

Norway rats. They become sexually mature at 3 to 5 months of age. Females can produce six litters, having six to eight young per litter. The average life span is 9 to 12 months.

Roof rats prefer to eat vegetable matter, including fruits, nuts, grains, and vegetables. Under stress, however, they eat a much wider array of foods.

This species less frequently nests in burrows in the ground and, due to its excellent ability to climb, often can be found nesting in vines, trees, and other types of dense foliage; occasionally they are found in sewers. In buildings, they nest in wall voids, attics, rafters, and other secluded and elevated locations. They are good climbers and can swim and squeeze through openings as small as ½ inch in diameter, therefore they can be difficult to exclude from buildings.

MICE

Many species of mice are pests, although the house mouse is the most common in buildings.

House Mouse
Mus musculus

Many varieties of house mouse (Figure 10-4) occur in the United States. Although they are pests in buildings, they also live outdoors. An adult house

FIGURE 10-4.

House mouse, Mus musculus.

mouse is about 3½ inches long, with a tail about the same length. They are usually dusky gray but may range from light brown to dark gray and commonly have a lighter underbelly. House mice have large, distinct ears. Adults weigh between ½ and 1 ounce. They reach sexual maturity within 35 days after birth. Gestation takes 18 to 21 days; mature females can produce a litter every 50 days, with an average of six young per litter. After about 15 months of age, females stop having litters. Both males and females may live several years.

House mice cause structural damage to buildings from gnawing and nest-building activities. They can damage attic and wall insulation and may also chew through electrical wiring. If they build nests in large appliances, they may destroy insulation and wiring. Through feeding and nesting, mice ruin items stored in warehouses, storerooms, attics, basements, and garages, and also seriously damage artifacts and collections in museums.

Like rats, house mice are capable of harboring several diseases. Diseases attributed to house mice include salmonellosis, rickettsialpox, and leptospirosis.

House mice eat most human food items. They consume meats, grains, cereals, seeds, fruits, and vegetables. A single mouse is capable of eating up to 8 pounds of food per year, although it destroys much more than this due to fecal and urine contamination or partial eating. Mice also damage food packaging materials and containers. They can go for long periods without water, although in locations where water is scarce they are attracted to fruits and other foods with a high water content. They require more water when high-protein food is consumed.

House mice climb well, are good swimmers, and can jump more than 12 inches. They are capable of crawling through openings as small as ¼ inch in diameter. They run easily along horizontal pipes, wires, beams, and other objects. House mice adjust rapidly to changes in their environment and explore new objects and try new food within a few hours after it is put out. They usually range an average of only 10 to 12 feet from their nest for food and water; at a maximum their travel is usually within a range of 30 feet, although they travel farther in some situations.

Management Guidelines for Rats and Mice

The principles of managing rats and mice are similar. However, it is important to identify the species so that control efforts can be tailored to its special habits. For more successful management, use several approaches including sanitation, exclusion, mechanical control, and chemical control.

Successful rodent control may be followed by an outbreak of secondary pests such as rat fleas or mites. Be sure to look for possible secondary pest problems and be prepared to use control measures for these pests while conducting a rodent control program.

Because rodents may be diseased or infested with parasites such as fleas or mites, wear gloves when handling and disposing of carcasses. Place carcasses in sealable plastic bags and dispose of them by burning or burying. Keep children and pets away from living or dead rodents.

Detecting and Monitoring

Rat and mice infestations can be detected by their feces and urine odors. Other identifying characteristics are holes and gnaw marks on structural

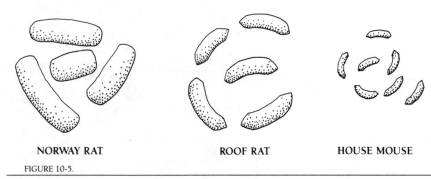

NORWAY RAT   ROOF RAT   HOUSE MOUSE

FIGURE 10-5.

The shape of fecal pellets can be helpful in identifying the species of rodent infesting a building.

portions of the building and trails and greasy markings along runways. Figure 10-5 compares fecal pellets of roof rats, Norway rats, and house mice; shape is more characteristic than size as younger rats produce smaller pellets than adults. Use a blacklight (ultraviolet light source) to see urine droplets and help locate areas of greatest and most recent rodent activity. Use these signs to help you determine how the rodents are getting into the building and how they are moving about the premises once inside. Search for areas where nests are located. For rats, stuff paper wadding into burrow entrances and check them in 24 hours to see if wadding has been removed or chewed, indicating active burrows. Determine what type of food is available to the rats or mice and where and how they are getting it. Look for water sources.

Use Table 10-1 to estimate the size of the rodent population. If possible, evaluate the extent of damage and economic loss caused by the rats or mice in order to estimate the degree of effort and amount of money that should be spent to eliminate the infestation. From this information, begin to develop a control strategy. Prepare a sketch of the floor plan of the building and indicate entry points, feeding areas, water supplies, runs, and nest sites. Begin by focusing control efforts in areas of greatest and most recent activity.

Exclusion

Exclude rodents from buildings by sealing or blocking all points of entry. Use durable materials such as concrete or metal for blocking holes to make it impossible for rodents to reopen them. Blocking entry routes should be included with all other methods of control, for reinvasion will quickly replace those animals destroyed by control efforts, especially if food and water are still available for them. See Table 2-4 on page 19 for information on materials used for excluding rodents. Be careful not to seal rodents into wall spaces or attics as they may die without food or water, and serious odor and fly problems can result.

Total exclusion may be impossible due to the size or design of a structure. Under these circumstances, use control methods aimed at reducing rodent populations around the outside of the building and reducing access to entry points. For instance, wrap rat guards (bands of thin sheet metal at least 18 inches wide) around the trunks of all trees adjacent to the building to keep rats from climbing up (Figure 10-6). Trim or remove dense foliage and ivy in contact with the building, and use cone-shaped shields on electrical and telephone wires coming to the building.

TABLE 10-1

Ways to Estimate the Size of a Rodent Population.*

POPULATION SIZE	VISUAL OBSERVATIONS
Rodents not present or in very low numbers. Any infestation is probably very recent.	None of the signs listed below have been observed.
Medium population	Old droppings present. Signs of gnawing seen. One or more rodents seen at night by flashlight. No rodents reported observed in the day. (There are probably 10 or more rodents in the general area where one is seen at night).
High population	Fresh droppings. Signs of recent gnawing. Tracks observed in dust. Three or more rodents seen at night by flashlight or one or more seen in daylight.

SIGNS INDICATING PRESENCE OF RODENTS

When searching for the presence of rodents in a building, look for these signs:

Sounds	Due to gnawing, feeding, fighting or moving about.
Droppings	Found along runways near nests, in feeding areas.
Urine	May be wet or dry; fluoresces in dark area .
Smudge marks	Found around pipes and beams and other structural parts of building.
Runs	Found next to walls, along fences, and under shrubbery. In buildings, runs may be dust-feee trails on floors, cabinets, and structural parts of building.
Gnawing	Wood chips, torn fabrics, tooth marks, damaged doors, door frames, window frames, moldings, cabinets, furniture, and other objects.
Visual sightings	Use flashlight or spotlight at night. Seeing rodents in daylight indicates a high population.
Nests and food caches	Found in attic and undisturbed areas, including dense shrubbery.
Pet excitement	Cats or dogs may probe area of floor or walls where rodents are nesting.
Odors	Usually can distinguish between rat and mouse odors.

*From Rat Control Manual by W. E. Howard and R. G. Marsh, published in *Pest Control*, volume 42-8, 1974.

Habitat Modification and Sanitation

Modify the environment inside and around the outside of the building to discourage rodents. Start by eliminating food and water supplies whenever possible. Areas where food is prepared or served must be cleaned daily to remove any food traces that can sustain a rodent population. All food must be stored in rodentproof containers. Establish policies for good housekeeping practices in other areas of the building to eliminate nest sites and materials that can be used for nesting. Outdoors, keep shrubbery and grass well trimmed and get rid of weeds and dense foliage that can provide nesting sites. Remove

FIGURE 10-6.

Barriers such as those shown here can be used to keep rats out of trees and gaining access to buildings.

trash heaps and other stored materials that provide nesting sites. Keep garbage containers tightly covered.

Trapping

Whenever possible, use mechanical or sticky traps for control of rodents inside a building. Figure 10-7 illustrates the different types of traps available for rat and mouse control; these include snap traps, glue boards, and live traps. Because they are nontoxic, traps are one of the safest methods that can be used for eliminating rats and mice. Trapping may produce quicker results than other control methods. Dead animals caught in traps can be located and

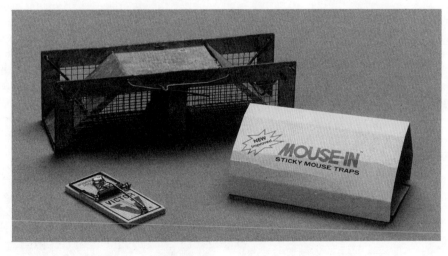

SPRING   LIVE   STICKY

FIGURE 10-7.

Rodents may be trapped by using snap traps, live traps, or glue boards.

disposed of more easily than those that are killed by poisons. Some individuals learn to avoid traps, however, and may become trap shy.

To be most effective, place traps along normal runways with triggers of spring traps placed adjacent to walls or other objects in the line of travel. If traps are used in areas where they are accessible to children or pets, put them in large bait boxes or similar containers. Trapping requires considerably more time than chemical control methods because traps must be checked and serviced daily.

Snap Traps. Snap traps are illustrated in Figure 10-8. In many cases, snap traps provide an effective method of controlling rodents without the need for rodenticides. They are useful in situations where there is a chance that poisoned animals might go into wall voids or other inaccessible areas to die. Other criteria for the use of snap traps include (1) if only a few animals are present, (2) in situations where people are opposed to the use of toxicants, and (3) when you can afford the investment of time needed for a successful trapping program.

To increase their effectiveness, place snap traps where rats or mice are most likely to encounter them, such as along or near known travel routes on floors, shelves, and walls. Lay a board against the wall as illustrated in Figure 10-9 to create a runway. Always use many traps to increase chances of target rodents being caught. Place two traps end to end as shown in Figure 10-10 to prevent rodents from jumping over and not being caught. Locate traps in areas inaccessible to children or pets. Also, do not put traps in locations where they might be accidentally tripped. In homes and offices, place traps out of sight so trapped animals will not be visible. Check traps daily to remove captured animals.

Rodent odors from previous catches do not repel other rodents and may even enhance the effectiveness of spring or live traps. Petroleum oils repel rodents, however, so never use these as a lubricant or to prevent rusting. If rusting is a problem, cover metal surfaces with lard or other animal fat.

To reduce trap shyness, bait the traps for several days without setting them until bait is being taken regularly; then set the spring mechanism. Bait snap traps with an attractive food or nesting material. Foods attractive to rodents include peanut butter, bacon, sausage, whole peanuts, chocolate candy, marshmallows, cheese, and dog or cat kibbles; many people have discovered other materials that work well in their situation. When unsure of which bait to use, try several different types at first, then continue baiting with the one that appears to work best. Tie bait securely to the trigger mechanism so it cannot be removed without springing the trap. Baits that cannot be tied, like peanut butter, have the disadvantage of being easily removed from the trap without setting off the trigger mechanism. Therefore, place only small amounts onto the trigger plate; the rodent must work harder to remove the bait so its chances of getting caught are increased. Replace bait frequently to prevent baited traps from losing their effectiveness as the bait becomes old or stale.

Baits used in traps may attract ants or cockroaches; if these are a problem, use nonbaited traps or bait them with nesting materials. For example, use a small, securely attached cotton ball (mice collect cotton fibers for their nests). Traps are also available equipped with trigger pedals infused with a mouse-attracting scent that does not attract insects.

When using nonbaited traps, enlarge the trigger mechanism with a piece of thin cardboard as illustrated in Figure 10-11. Traps like these are available commercially. Place nonbaited traps directly in rodent pathways, close to walls or other vertical surfaces; position the trigger side of the trap toward the wall.

FIGURE 10-8.

Snap traps are often an effective method of controlling rats or mice, eliminating the need for using rodenticides.

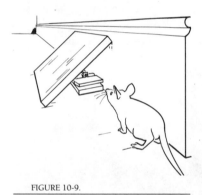

FIGURE 10-9.

Place snap traps along walls and cover them with a board. This will force rodents to walk over the trap.

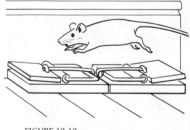

FIGURE 10-10.

For greater effectiveness, place traps in pairs along walls. This will prevent rodents from jumping over a trap and avoid being caught.

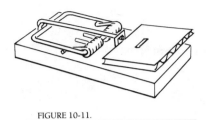

FIGURE 10-11.

Spring traps can be more effective and can be used without bait if the trigger pedal is enlarged using a piece of heavy cardboard. Traps can also be purchased with an enlarged trigger pedal.

FIGURE 10-12.

Glue boards, also known as sticky traps, are be used for trapping mice, rats, and insects such as cockroaches.

Also nail nonbaited traps to beams and other overhead passageways used by rats; be sure the trigger side of the trap is positioned in the line of travel.

Trapping rodents is most successful during the first few days after traps have been placed. After 3 or 4 days, rodent catches usually drop off. Once the initial trapping period has passed, use a nontoxic tracking powder to see where remaining rodent activity is taking place and concentrate trap placement in those areas.

Glue Boards. Use glue boards in the same manner as snap traps to catch mice and rats (Figure 10-12). Glue boards, also called sticky traps, are disposable cardboard or plastic units having one or more surfaces thickly coated with a sticky paste.

Locate glue boards in runways and tape them around pipes or other objects traversed by rodents. Once an animal becomes caught, dispose of the trap and pest. Live rodents may be killed by submersing the trap in water or placing it in a freezer. For control of mice, glue boards have similar advantages to snap traps. Keep them away from children or pets. They do not work well in dusty locations where sticky surfaces become coated with dust and debris. If dusty conditions exist, place glue boards in protected bait boxes and use nonpoisonous bait to attract rodents to the boxes.

Check glue board traps daily. Do not reuse traps once a rodent has been caught as some of the sticky substance will be lost and the trap is no longer as effective. Rats, being larger and stronger, are more difficult to capture on glue boards; secure traps to a surface with tape or tacks to keep rats from dragging them away.

Live Traps. Live traps can be used to capture rats and mice; they are also used for capturing birds, skunks, opossums, and other small animals (Figure 10-13). Live traps are the only type of traps that can be used for protected wildlife such as owls.

Live trapping also requires time and patience. Use a nontoxic food as a bait. Suitable baits include grains, fruits, meat, and other items. To increase trapping success, keep unset traps supplied with fresh bait for several days. Set the trap closing mechanism after food is being taken on a regular basis. An

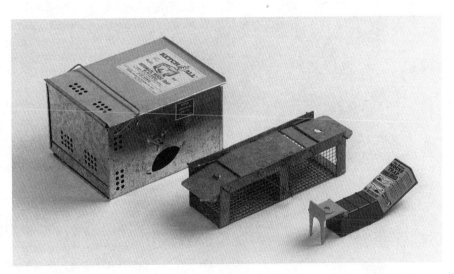

FIGURE 10-13.

Live traps can be used to capture or monitor the presence of small animals such as birds or rodents without injuring the animals.

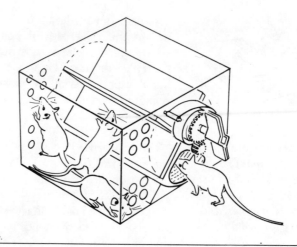

FIGURE 10-14.

This multiple catch trap will hold up to 15 mice and does not have to be reset each time one is caught.

animal that has escaped from a trap will probably not go into one again, so do a careful, thorough job of trapping the first time.

Place live traps in areas where they will not be disturbed by children or pets and where they will not capture cats or small dogs. Check traps daily or more frequently to make sure that bait is available and that the trap is still set.

Several styles of live traps are available for catching mice. A simple plastic one has a door that snaps shut because the trap tips as a mouse enters to get the bait. A more complicated mechanical live trap for mice is powered by a spring-wound mechanism; when a mouse enters the baited or unbaited entrance it trips a lever, causing it to be flipped into a holding chamber (Figure 10-14). This trap holds several mice and does not have to be reset each time one is caught.

Rodenticides

Rodenticides are divided into three groups based on their mode of action or method of application. These groups are (1) multiple-dose and single-dose anticoagulants, (2) acute single dose non-anticoagulant toxicants, and (3) fumigants. Anticoagulants and acute toxicants may be applied as tracking powders, food baits, or liquid baits. Bait shyness and resistance problems may be associated with careless use of rodenticides, so keep bait fresh, use it selectively, and use it only when nonchemical methods are not suitable. Rodenticide baits must be carefully placed to ensure that target rodents walk through or consume the toxic material. For instance, placement is different for control of roof rats (in trees, etc.) than for Norway rats (on the ground) because these two species prefer to nest and feed in different environments.

An understanding of the rodent's habits is required to successfully use rodenticides. Rats, for example, are wary of new items in their environment and may take several days before investigating or tasting introduced bait. Mice, on the other hand, usually investigate new items within a few hours.

Baits. Poisonous baits are effective ways of controlling rodents in a confined area. Use them especially if access to other sources of food or water can be reduced or eliminated. Baits must be attractive to the target rodents, therefore they should be some form of a natural or acceptable food. Sometimes the attractiveness of rodent baits can be enhanced by combining them with sugar, molasses, corn syrup, or similar sweetener or with vegetable oil or animal fat.

Keep baits out of the reach of children or pets. Rodenticides are one of the leading causes of dog poisonings. These materials may also attract ants, stored-product pests, or other insects.

Some baits are available as bait blocks, which can be attached to walls or structures in areas where rodents are most active. Most bait blocks incorporate paraffin or other waxy materials to keep them fresh and help protect them against moisture. Dogs may chew on paraffin blocks, so put them in out-of-reach locations.

Baits in the form of grains or pellets should be placed in bait stations so the toxic material will not become scattered. Locate bait stations along known travel routes and near nest sites. Record the location of each bait station within a treated area so that each station can be properly maintained and so all stations will be removed when the baiting program is terminated. Be sure bait stations are marked to indicate that they contain poisonous bait; also attach a label that includes the signal word, chemical name, and the name and telephone number of the person responsible for the bait station. The words *"keep out of reach of children"* should be clearly printed on the bait station.

Prebaiting for several days with untreated bait may be useful as a monitoring technique and may help to determine how much toxic bait to use in a bait station. Once the untreated bait is being consumed regularly, switch to a bait of the same type that has been treated with a toxicant. In this way rodents should become accustomed to the bait stations and associate them as a food source. Whether using treated or untreated bait, check the bait stations frequently to be sure there is an ample supply of fresh bait. If bait is not being taken in some stations but is being eaten from others, consider relocating the unused stations. If little or no bait is being removed from all the bait stations, the bait may not be suitable or there may be another food source that is more attractive. Bait stations may also be located in the wrong places. First, if possible, eliminate all competing food sources. Be sure bait stations are positioned near nests or known travel routes. If bait acceptance is still poor, switch to another type or brand of bait. Should this fail, use other methods of control such as toxic tracking powder and trapping.

Fumigants. Most fumigants used for rodent control are injected or released into the burrows of the pest or are used in food storage areas such as grain silos. A disadvantage to fumigation is that dead rodents trapped in inaccessible sites may result in serious odor and fly problems.

Fumigation requires that the treated area be evacuated and sealed so that the airborne concentration of toxicant can reach a lethal level. The toxicant concentration must be held at this level for a specific period of time as described on the label. Afterward, the area must be thoroughly ventilated before it is safe to enter. Refer to the supplement to this publication entitled *Fumigation Practices* for information on how to use fumigants and how to prepare an area for fumigation.

BATS

Bats, among the most numerous land vertebrates, occur in about 2000 species worldwide; 23 species are known in California (Figure 10-15). Bats are nocturnal flying mammals, and all but one California species feeds on insects. The remaining species visits flowers to feed on nectar.

FIGURE 10-15.

Bats may roost in buildings, causing odors and harboring pests and disease organisms.

During the day, bats seek shelter in dark, protected areas; they prefer locations with fairly high humidity. Once an appropriate site is found, the shelter may become a long-term roost site. Although individuals of some species live singly (free living), many species congregate into colonies, some of which have greater than 1 million individuals. Bats commonly roost in or near buildings, which can cause problems such as noise, smell, accumulations of feces (guano) and urine, staining and spotting of surfaces, associated parasites, and potential for disease transmission. They may occasionally attract flies and cockroaches. In addition, there is a general fear of these mammals by the public. Migratory species inhabit buildings only during certain times of the year, but other species may be present in the same location year around. Roost locations include attics, wall voids, hollow floors, chimneys, unused furnaces, belfries, and decorative structures.

Most species of bats occurring in California range from 3 to about 5½ inches in length and may be various shades of gray or brown. Females of smaller species usually produce a single offspring each year during late spring to early summer. When first born, young bats cling to their mothers. As they grow and become too heavy to carry, they are left at the roosting site while females forage for food; offspring are weaned when they are able to fly and begin to forage for themselves. Adult bats may each catch and consume more than 500 insects per hour, and are therefore beneficial in this respect.

Management Guidelines for Bats

Bat identification is helpful in planning a control program. Identification provides access to information on habits and biology. For example, species that migrate may not require control because they may only be in a building temporarily. Solitary species are much less of a problem than species that live in large colonies.

Bats are protected under various state or local regulations. Check with wildlife authorities before planning a management program to determine what control methods are acceptable; *killing of bats is prohibited in California* except by special authorization, and then only if they are causing damage to property. The single most important way of controlling bats in structures is to exclude them, but only attempt to do this before young have been born or after they are old enough to leave the roost to forage; never attempt control between mid-May and mid-August as young will be trapped inside, where they will die of starvation and create odor problems.

To begin the exclusion, locate all openings into the building used by the bats for their nightly departures. Many species are able to squeeze through openings as small as ⅜ inch in diameter. Some openings can be located by looking for smudges or fecal droppings; however, it may be necessary to observe bats leaving or arriving during the hours after dusk and before dawn to make sure each opening is located. While bats are at roost, close off all but one or two main openings. Use caulking, wood, sheet metal, plaster, cement, or ¼-inch mesh hardware cloth or netting to close entrances. Bats do not chew, so almost any type of sturdy material can be used. Close remaining openings during the night when all bats have left the roost to forage; be sure no young or adults remain in the roost area. If some bats have been trapped during the first night, reopen a few exit holes the following evening to allow their escape. Figure 10-16 shows a design for an entrance covering that allows bats to leave but prevents their reentry. If entrances are successfully blocked, bats are unable to return to this roosting area and will move to another location. Control of migratory species can be accomplished in the same manner except that openings should be blocked during the period after the bats have migrated; when they return the following year, they will be unable to enter the building. Migratory bats leave for warmer climates during fall and winter months.

Several other management techniques can be used to control bats. However, unless all openings are blocked, there is always the possibility of the same or different individuals returning to the roosting site. For example, odors from excrement of roosting bats attract other bats to the area. Installing bright lights sometimes helps to discourage roosting and can be a useful technique in warehouses or similar buildings that are difficult to keep closed up. Fans

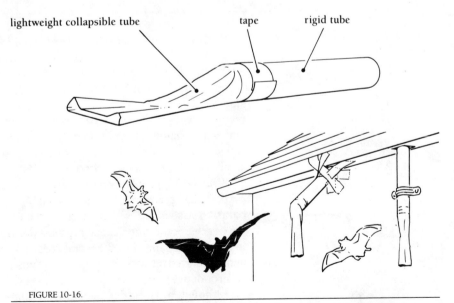

lightweight collapsible tube   tape   rigid tube

FIGURE 10-16.

Blocking entrance holes with cloth, netting, or the device shown in this illustration allows bats to leave a building but prevents them from getting back in.

that blow directly on the roosting bats sometimes help repel them unless they are able to find sheltered areas protected from the moving air. High-frequency whistles have been shown to be effective in repelling bats. Fiberglass insulation also discourages roosting; it is suspected that the small glass fibers irritate the bat's skin.

When their numbers are few, bats can sometimes be caught with the aid of traps or nets, then removed from the building and released. Be careful when handling bats because they have powerful jaws and extremely sharp teeth with which they can inflict a painful bite. Bats may be infected with the rabies virus, which along with other diseases can be transmitted by their bite. Estimates on the incidence of rabies have ranged from 1 out of every 1000 to 1 out of every 10 bats in a colony. If bitten or scratched by a bat, cleanse the wound thoroughly with soap and water and obtain prompt medical advice. Try to collect the offending bat so it can be tested for rabies. Do not handle bats, even if you are wearing heavy protective gloves.

Bats may have one or more species of external parasites that build up to large numbers in roosting sites. Parasites include bat bugs, bed bugs, ticks, mites, and fleas. Apply desiccant dusts throughout the roosting site to kill parasites before they spread and locate human hosts once the bats are gone.

Bat guano dries to form a crumbly, powdery substance that is capable of supporting the growth of a fungal organism that causes a systemic disease (known as histoplasmosis) in people. Effects of this disease range from flulike symptoms to serious lung abscesses and lesions resembling tuberculosis.

Fresh bat urine serves as a medium for possible transmission of disease organisms from an infected bat. When working in an area where bats are roosting, wear protective clothing and a cartridge respirator to avoid skin contact with urine or breathing guano dust. Never handle live or dead bats. Thoroughly wash reuseable protective clothing to prevent spreading fungal spores, or wear disposable coveralls.

No pesticides are registered for bat control in California.

FIGURE 10-17.

Birds can sometimes be pests in and around buildings because they are noisy, create messes, and harbor parasites that may attack people.

BIRDS

Birds sometimes cause problems in and around buildings (Figure 10-17). They are often noisy and highly visible animals that attract considerable attention. Birds are generally regarded as desirable components of the environment, so much so that some control programs are publicly challenged. The sight of dead birds can set off a considerable public outcry, requiring that bird management programs be designed to be sensitive to the public's feelings while at the same time eliminating damage or other problems that make birds a pest. Many bird species are protected by state and federal wildlife laws or local ordinances which restrict methods that can be used for their control.

Birds create problems in and around buildings because of their noise, droppings, feeding, mating, and nesting habits. They also have several external parasites associated with their nests which may attack people working or living nearby. Some birds cause physical damage to buildings by pecking holes in roofs, eaves, and siding. Several species live and nest inside large warehouses and similar structures. Besides being nuisances, birds are serious hazards to aircraft at airports.

Although most birds have the potential for becoming pests, pigeons, starlings, house sparrows, cliff swallows, flickers, and woodpeckers are the species typically considered pests. Refer to Table 10-2 for identification keys and helpful information on these bird species.

TABLE 10-2

*Identification and Management of Pest Birds.**

SPECIES	DESCRIPTION	BEHAVIOR	DAMAGE
BLACKBIRD	Many species; 6 to 16 inches; females smaller bodied; sharp, pointed bills; plumage iridescent black; some species have brightly colored areas of yellow, red or orange on head or wings; female plumage brownish, often with streaked breast.	Gregarious; flock ranges from few birds to thousands; some species congregate in huge winter roosts.	Eats vegetables (lettuce, peppers, tomatoes, sweet corn), and nuts (sunflowers, almonds).
CROWNED SPARROW	Two species, ranging from 5¾ to 7 inches; typical sparrow coloration; brownish on back, dull, grayish breast. Adult white-crowned sparrow: three white strips on crown. Adult golden-crowned sparrow: dull gold crown edged with black.	Forages on ground in grassy and open areas near brush, fencerows, and other such cover.	Feeds on vegetable and fruit crops, especially lettuce, grapes, melons, almonds, strawberries; disbuds fruit and nut trees; damages young seedlings in fall and winter.
GOLDFINCH	Two species: American goldfinch, 4½ inches; male has bright yellow back and breast, black cap and wings; in winter resembles duller-colored female. Lesser goldfinch, 3¾ inches; male has dark head and back, bright yellow breast; female is dull-colored with dark wings.	Lives in small flocks in weedy fields, brushes, and roadsides; swoops up and down in flight.	Eats flower and vegetable seeds, strawberries and sunflowers; disbuds almonds and apricots.
HORNED LARK	6½ to 7 inches; light brown body above, black band across breast, black strip from bill to eyes; two black "horns" above eyes; walks with slight sideways swaying of body and fore-and-aft movement of head; does not hop on ground.	Ground bird found in loose flocks in wide, sparsely vegetated open areas; normally flies low, and swoops up and down slightly.	Feeds on vegetables (lettuce, broccoli, carrots), melons, flowers, particularly at seeding stage.
HOUSE FINCH	5 to 5¾ inches; male has rosy-red head, rump, and breast, brownish back and wings, sides streaked with brown; female lacks red, has brownish body with heavily streaked breast and abdomen.	Well adapted to human environments, often nests in vines on buildings; sings, "chirps" from trees, antennas, or posts; found in variety of habitats, from deserts and open woods to farmlands, suburbs, and farms.	Eats fruits and berries in orchard and garden; disbuds and deflowers fruit and nut trees; attacks seed crops.

	SPECIES	DESCRIPTION	BEHAVIOR	DAMAGE
	HOUSE SPARROW	5¾ to 6¼ inches; male has black bib and bill, white cheeks and gray cap; female is dull brown above and dingy whitish below without black bib, bill, or gray cap.	Abundant on farms, in cities and suburbs; lives in loose flocks; often nests in buildings eaves, vents, or other openings and cavities.	Eats emerging seedlings, fruit, buds; damages flowers, newly seeded lawns, ripening fruit; droppings deface buildings.
	MAGPIE	Large bird, 16 to 20 inches long; black and white body with long, streaming tail.	Lives in farming areas of California valleys and nearby foothills; gregarious, found in colonies; builds large stick nest high in trees near open grasslands or fields.	Feeds on fruit, nuts, grain, garbage.
	PIGEON	14 to 15 inches; plump-bodied, short-billed; usually blue-gray with whitish rump and red feet, but white, brown, or other-colored plumage not uncommon.	Found in cities and suburbs; feeds on seeds, grain, fruits, insects; coos intermittently while perched; roosts in large flocks.	Deposits droppings on buildings and cars, contaminates foodstuffs; nests on buildings, may clog drain pipes; transmits disease to humans and domestic animals.
	SCRUB JAY	10 to 12 inches; head, wings, and tail blue; underparts and back gray; white throat; no crest.	Found throughout California except in deserts and high mountains; very noisy, makes short flights ending in sweeping glide.	Eats orchard fruits and nuts.
	STARLING	7½ to 8½ inches; short tail; long, slender, yellow bill in spring and summer, dark bill in winter; plumage black to purplish-black; heavily speckled in winter.	Abundant in city parks, suburbs, and on farms; gregarious; uses large communal roosts from late summer until spring; flies swiftly and directly; primarily ground feeder.	Pulls small plants; damages fruit (grapes, cherries, strawberries, and others); nests in building eaves and other openings; droppings deface buildings.
	WOODPECKER (FLICKER)	Several species in California, size varies from 5¾ to 15 inches; all have strong, sharp-pointed bill for chipping and digging in tree trunks and branches for insects; uses stiff tail as a prop. One species, the flicker, is jay-sized woodpecker with brown back, white rump, usually salmon-red under wings, but occasionally yellow.	Most species peck or "drum" repeatedly on resonant limbs, poles, or drainpipes; usually undulate in flight, folding wings against the body after each series of flaps. All species nest in excavated holes. Flickers are often seen on ground eating ants.	Woodpeckers do structural damage, drilling into siding and shingles and under eaves of buildings for food or to excavate nest chamber; damages fences, poles. Drumming on buildings may create annoying noise.

*From *Wildlife Pest Control Around Gardens and Homes* by Terrell P. Salmon and Robert E. Lickliter. UCANR Publication 21385.

Management Guidelines for Pest Birds

Because many species of birds are protected by federal and state wildlife laws or by local regulations, it is necessary to identify the pest birds and the nonpest species associated with them that may be affected by control efforts. For control restrictions and management information, check with the regional offices of the California Department of Fish and Game, the U.S. Fish and Wildlife Service, or the local agricultural commissioner (see Table 10-3).

Evaluate bird problems carefully and determine what reasonable goals should be expected by a management program. Find out what attracts birds to a particular building or area—it could be suitable roosting and nesting sites, food supplies, or water needed for drinking or bathing.

If possible, modify the building or surrounding area to discourage bird roosting or nesting. Build up flat areas so they slope and are not suitable for nests or install commercially available metal strips that have pieces of wire protruding in several directions. Use these metal strips or plastic or wire netting to keep birds away from eaves or other nesting or roosting sites. Exclude birds from large buildings such as warehouses by placing netting over entrances. Use netting to keep birds out of certain outdoor areas, such as court-yards, adjacent to buildings. Close holes and crevices with wood or metal to keep these from being used as nesting sites. Eliminate food and water sources, if possible, that attract birds. Food sources include garbage, grain, fruit trees, berry plants, ornamental fruiting plants, seeds, and insects.

Live trapping may be a successful method of controlling certain species of birds around buildings. Live traps are baited with suitable food such as seeds.

Frightening devices such as artificial snakes or models of predatory birds have limited use in keeping birds away from buildings. These lose their effectiveness quickly once birds become accustomed to them. Metallic or brightly colored strips attached to wires that move with air currents may help to discourage birds from roosting in some locations, although this technique is not usually effective after a few days.

Several companies manufacture sound-generating devices for scaring away birds. These devices may have only limited use around buildings because the sounds they generate can also be annoying to people living or working in the area. Sound-generating devices are only effective over a limited distance, so several units may be required. Birds may become accustomed to sounds generated by these devices after a short period of time; relocating the sound generators from time to time helps reduce this problem.

Birds do not have an acute sense of smell, so are therefore not affected by olfactory repellents that can be effective with other types of vertebrates. In some limited situations, however, sticky substances can be spread on railings and other surfaces used as perches or roosts to repel birds.

Birds, such as pigeons, may create a collection of fecal material where they roost or loaf. Often this material must be removed. During removal, wear protective clothing and respiratory protection to reduce exposure to the *Histoplasma* fungus responsible for the disease known as histoplasmosis that can affect the lungs, spleen, central nervous system, and other internal organs in people.

TABLE 10-3

Locations of the U.S. Fish and Wildlife Service Endangered Species Offices and the Local and Regional Offices of the California Department of Fish and Game.

U.S. FISH AND WILDLIFE SERVICE
Federal Building
24000 Avila Road
Laguna Niguel, CA 92656
(714) 643-4270

This office serves the counties of Imperial, Los Angeles, Orange, Riverside, San Bernardino, San Diego, Santa Barbara, and Ventura

U.S. FISH AND WILDLIFE SERVICE
Sacramento Endangered Species Office
2800 Cottage Way, Room E-1823
Sacramento, CA 95825
(916) 978-4866

This office serves all the remaining counties in California

CALIFORNIA DEPARTMENT OF FISH AND GAME
1416 Ninth Street, 12th Floor
Sacramento, CA 95814
(916) 445-3531

REGIONAL OFFICES

REGION I
601 Locust Street
Redding, CA 96001
(916) 225-2300

REGION IV
1234 E. Shaw Avenue
Fresno, CA 93710
(209) 222-3761

REGION II
1701 Nimbus Road
Rancho Cordova, CA 95670
(916) 355-0978

REGION V
330 Golden Shore, Suite 50
Long Beach, CA 90802
(213) 590-5126

REGION III
7329 Silverado Trail
Napa, CA 94558
  Mailing Address:
    P.O. Box 47
    Yountville, CA 94599
(707) 944-5500

MARINE RESOURCES REGION
330 Golden Shore, Suite 50
Long Beach, CA 90802
(213) 590-5189

BRANCH OFFICES THAT CAN ALSO PROVIDE INFORMATION

407 Westline Street
Bishop, CA 93514
(619) 872-1171

2201 Garden Road
Monterey, CA 93940
(408) 649-2870

619 Second Street
Eureka, CA 95501
(707) 445-6493

1350 Front Street
Room 2041
San Diego, CA 92101
(619) 237-7311

References

Barker, R. L., and G. C. Coletta, Eds. 1986. *Performance of Protective Clothing*. ASTM Special Technical Publication 900, Philadelphia, PA

Bond, E. J. 1984. *Manual of Fumigation for Insect Control*. Food and Agriculture Organization of the United Nations, Rome, Italy

California Department of Food and Agriculture 1980. *Laws and Regulations Study Guide for Agricultural Pest Control Adviser, Agricultural Pest Control Operator, Pesticide Dealer, and Pest Control Aircraft Pilot Examinations*. CDFA, Sacramento, CA

Daar, S., and W. Olkowski. "Moisture Management: Key to Protecting Your Home." *Common Sense Pest Control* I(4) (1985): 13-21.

Ebeling, W. 1975. *Urban Entomology*. University of California Publication 4057, Berkeley, CA

Edwards, S. R., B. M. Bell, and M. E. King 1981. *Pest Control in Museums: A Status Report (1980)*. Association of Systematics Collections, Lawrence, KS

Kofoid, C. A., Ed. 1934. *Termites and Termite Control*. University of California Press, Berkeley, CA

Mallis, A. 1982. *Handbook of Pest Control*, 6th Edition. Franzak and Foster Company, Cleveland, OH

Morgan, D. P. 1989. *Recognition and Management of Pesticide Poisonings*, 4th Edition. U. S. Environmental Protection Agency, Office of Pesticide Programs, Washington, D.C.

Olkowski, H. "Carpet and Hide Beetles and What to do About Them." *Common Sense Pest Control* IV(1) (1988): 5-10.

Olkowski, H., and W. Olkowski. "The Argentine Ant: Pest and Predator." *Common Sense Pest Control* V(1) (1989): 13-18.

Olkowski, H., and W. Olkowski. "Ants in the House." *Common Sense Pest Control* IV(4) (1988): 6-16.

Olkowski, W., and S. Daar. "Pantry Pests: Beetles and Moths in Stored Foods." *Common Sense Pest Control* II(4) (1986): 16-19.

Olkowski, W., and H. Olkowski. "Safe Ways to Remove Bats from Buildings." *Common Sense Pest Control* V(3) (1989): 5-13.

Olkowski, W., and H. Olkowski. "The Why of Flies: Common Sense Management of Flies Associated with Garbage and Manure." *Common Sense Pest Control* II(4) (1986): 4-13.

Olkowski, W., and H. Olkowski. "Clothes Moths: How to Protect Your Woolens." *Common Sense Pest Control* II(2) (1986): 7-12.

Olkowski, W., and H. Olkowski. "Carpenter Ants." *Common Sense Pest Control* I(2) (1985): 11-16.

Olkowski, W., and H. Olkowski. "Carpenter Bees." *Common Sense Pest Control* I(2) (1985): 17-19.

Olkowski, W., H. Olkowski, and S. Daar. "Least-Toxic Pest Management for Fleas." *Common Sense Pest Control* VI(3) (1990): 6-14.

Olkowski, W., H. Olkowski, and S. Daar. "Managing Yellowjacket Pests." *Common Sense Pest Control* I(3) (1985): 4-11.

Olkowski, W., H. Olkowski, and S. Daar. "Controlling Wood-Boring Beetles in Buildings." *Common Sense Pest Control* I(3) (1985): 11-17.

Swan, L. A., and C. S. Papp 1972. *The Common Insects of North America.* Harper and Row, New York, NY

Truman, L.C., G. W. Bennett, and W. L. Butts 1982. *Scientific Guide to Pest Control Operations,* 3rd Edition. Harcourt Brace Javanovich, Duluth, MN

University of California Publications:
A Key to Ants of California. 1987. Publication 21433.
Carpet Beetles and Clothes Moths. Rev. 1979. Publication 2524.
Cliff Swallows: How to Live with Them. 1981. Publication 21264.
Common Pantry Pests and Their Control. Rev. 1977. Publication 2711.
Controlling Ground Squirrels around Structures, Gardens, and Small Farms. 1980. Publication 21179.
Controlling Household Cockroaches. Rev. 1982. Publication 21035.
Fate of Pesticides in the Environment. 1987. Publication 3320.
Guide to Vertebrate Pest Control Materials Registered in California. 1981. Publication 21226.
How to Control Bats in Your Home. Rev. 1984. Publication 2696.
Pesticides in Soil and Groundwater. 1983. Publication 3300.
Powderpost Beetles and Their Control. Rev. 1980. Publication 21017.
Silverfish and Firebrats: How to Control Them. 1977. Publication 21001.
Termites and Other Wood-Destroying Insects. Rev. 1990. Publication 2532.
The House Mouse: Its Biology and Control. 1981. Publication 2945.
The Illustrated Guide to Pesticide Safety. 1991. Publication 21488, 21489.
The Safe and Effective Use of Pesticides. 1988. Publication 3324.
Toxicology: The Science of Poisons. 1981. Publication 21221.
The Rat: Its Biology and Control. Rev. 1981. Publication 2896.
Wildlife Pest Control around Gardens and Homes. 1984. Publication 21385.

Glossary

abiotic. nonliving factors, such as wind, water, temperature, or soil type or texture.

absorb. to soak up or take in a liquid or powder.

acaricide. a pesticide used to control mites.

accumulate. to increase in quantity within an area, such as the soil or tissues of a plant or animal.

active ingredient (a.i.). the material in the pesticide formulation that actually destroys the target pest or performs the desired function.

adsorb. to take up and hold on surface.

aerosol. very fine liquid droplets or dust particles often emitted from a pressurized can or aerosol generating device.

agitator. a mechanical or hydraulic device that stirs the liquid in a spray tank to prevent the mixture from separating or settling.

annual. a type of plant that passes through its entire life cycle in one year or less.

antibiotic. a substance produced by a living organism, such as a fungus, that is toxic to other types of living organisms. Sometimes used as a pesticide.

anticoagulant. a type of rodenticide that causes death by preventing normal blood clotting.

arthropod. an animal having jointed appendages and an external skeleton, such as an insect, a spider, a mite, a crab, or a centipede.

attractant. a substance that attracts a specific species of animal to it. When manufactured to attract pests to traps or poisoned bait, attractants are considered to be pesticides.

attractant trap. a device use to monitor pests and pest activity. These usually contain a pheromone or food substance that attracts the pests and a sticky surface or some other method to trap the pest.

avicide. a pesticide used to control pest birds.

bacterium. a unicellular microscopic plantlike organism that lives in soil, water, organic matter, or the bodies of plants and animals. Some bacteria cause plant or animal diseases. (plural: bacteria).

bait. a food or foodlike substance that is used to attract and often poison pest animals.

beneficial. pertaining to being helpful in some way to people, such as a beneficial plant or insect.

biennial. a plant that completes part of its life cycle in one year and the remainder of its life cycle in the following year.

biochemical. pertaining to a chemical reaction that takes place within the cells or tissues of living organisms.

biological control. the action of parasites, predators, pathogens, or competitors in maintaining another organism's density at a lower average than would occur in the their absence. Biological control may occur naturally in the field or be the result of manipulation or introduction of biological control agents by people.

biotic. pertaining to living organisms, such as the influences living organisms have on pest populations.

blacklight trap. a monitoring or control device for certain flying insects; insects are attracted to the ultraviolet light source built into the trap.

botanical. derived from plants or plant parts.

broad-spectrum pesticide. a pesticide that is capable of controlling many different species or types of pests.

broadleaf. one of the major plant groups, known as dicots, with net veined leaves usually broader than grasses. Seedlings have two seed leaves (cotyledons); broadleaves includes many herbaceous plants, shrubs, and trees.

buffer area. a part of a pest infested area that is not treated with a pesticide to protect adjoining areas from pesticide hazards.

calibration. the process used to measure the output of pesticide application equipment so that the proper amount of pesticide can be applied to a given area.

California Department of Food and Agriculture (CDFA). the state agency responsible for regulating the use of pesticides in California.

carcinogenic. having the ability to produce cancer.

carrier. the liquid or powdered inert substance that is combined with the active ingredient in a pesticide formulation. May also apply to the water or oil that a pesticide is mixed with prior to application.

carrying capacity. the capacity a certain defined area has for supporting a population of pests; factors influencing a carrying capacity include food, water, temperature, light, humidity, and shelter or hiding places.

chronic. pertaining to long duration or frequent recurrence.

chronic onset. symptoms of pesticide poisoning that occur days, weeks, or months after the actual exposure.

common name. the recognized, nonscientific name given to plants or animals. The Weed Science Society of America and the Entomological Society of America publish lists of recognized common names. Many pesticides also have common names.

compatible. the condition in which two or more pesticides mix without unsatisfactory chemical or physical changes.

confined area. enclosed spaces such as attics, crawl spaces, closed rooms, warehouses, greenhouses, holds of ships, and other areas that may be treated with pesticides.

contact poison. a pesticide that provides control when target pests come in physical contact with it.

coverage. the degree to which a pesticide is distributed over a target surface.

danger. the signal word used on labels of pesticides in toxicity Category I—those pesticides with an oral LD_{50} less than 50 or a dermal LD_{50} less than 200 or those having specific, serious health or environmental hazards.

deactivation. the process by which the toxic action of a pesticide is reduced or eliminated by impurities in the spray tank, by water being used for mixing, or by biotic or abiotic factors in the environment.

degradation. the breakdown of a pesticide into an inactive or less active form. Environmental conditions, impurities, or microorganisms can contribute to the degradation of pesticides.

deposition. the placement of pesticides on target surfaces.

dermal. pertaining to the skin. One of the major ways pesticides can enter the body to possibly cause poisoning.

desiccant. a pesticide that destroys target pests by causing them to lose body moisture.

disease. a condition, caused by biotic or abiotic factors, that impairs some or all of the normal functions of a living organism.

dissolve. to pass into solution.

dose. the measured quantity of a pesticide. Often the size of the dose determines the degree of effectiveness, or, in the case of poisoning of nontarget organisms, the degree of injury.

drift. the movement of pesticide dust, spray, or vapor away from the application site.

dust. finely ground pesticide particles, sometimes combined with inert materials. Dusts are applied without mixing with water or other liquid.

economic damage. damage caused by pests to plants, animals, or other items which results in loss of income or a reduction of value.

economic threshold. the point at which the value of the damage caused by a pest exceeds the cost of controlling the pest, therefore it becomes practical to use the control method.

efficacy. the ability of a pesticide to produce a desired effect on a target organism.

emulsifiable concentrate. a pesticide formulation consisting of a petroleum-based liquid and emulsifiers that enable it to be mixed with water for application.

endangered species. rare or unusual living organisms whose existence is threatened by people's activities, including the use of some types of pesticides.

environment. all of the living organisms and nonliving features of a defined area.

Environmental Protection Agency (EPA). the federal agency responsible for regulating pesticide use in the United States.

eradication. the pest management strategy that attempts to eliminate all members of a pest species from a defined area.

evaporate. the process of a liquid turning into a gas or vapor.

exclusion. a pest management technique that uses physical or chemical barriers to prevent certain pests from getting into a defined area.

exposure. coming in contact with a pesticide.

flowable. a pesticide formulation of finely ground particles of insoluble active ingredient suspended in a petroleum-based liquid combined with emulsifiers; flowables are mixed with water for final application.

flypaper. strips of paper coated with a sticky substance and sometimes a pheromone attractant; these strips are hung in areas inside buildings where flies are a problem. Flies become entangled in the sticky substance.

fog. a spray of very small pesticide-laden droplets that remain suspended in the air.

formulation. a mixture of active ingredient combined during manufacture with inert materials. Inert materials are added to improve the mixing and handling qualities of a pesticide.

fumigant. vapor or gas form of a pesticide used to penetrate porous surfaces for control of soil dwelling pests or pests in enclosed areas or storage.

fungicide. a pesticide used for control of fungi.

fungus. multicellular lower plant lacking chlorophyll, such as a mold, mildew, rust, or smut. The fungus body normally consists of filamentous strands called the mycelium and reproduces through dispersal of spores (plural: fungi).

glue board. a small cardboard sheet or boxlike apparatus having one or more surfaces coated with a thick sticky paste. This is placed on surfaces to capture pest insects or small rodents.

granule. a dry formulation of pesticide active ingredient and inert materials compressed into small, pebblelike shapes.

groundwater. fresh water trapped in aquifers beneath the surface of the soil; one of the primary sources of water for drinking, irrigation, and manufacturing.

habitat. the place where plants or animals live and grow.

habitat modification. a pest management practice that involves modifying certain physical aspects of a building or structure to make it less suitable for pests to live.

haltere. a knoblike organ replacing the second pair of wings on flies, mosquitoes, and other insects in the order Diptera; this organ is believed to assist in balance.

herbaceous. a plant that is herblike, usually having little or no woody tissue.

herbicide. a pesticide used for the control of weeds.

host. a plant or animal species that provides sustenance for another organism.

host resistance. the ability of a host plant or animal to ward off or resist attack by pests or to be able to tolerate damage caused by pests.

impregnate. an item, such as a flea collar, that has been manufactured with a certain pesticide in it; impregnates usually emit small, localized quantities of pesticide over an extended period of time.

incompatibility. a condition in which two or more pesticides are unable to mix properly or one of the materials chemically alters the other to reduce its effectiveness or produce undesirable effects on the target.

incorporate. to move a pesticide below the surface of the soil by discing, tilling, or irrigation. To combine one pesticide with another.

inert dust. a finely ground clay or other powder used to control certain types of insects by desiccation.

inert ingredients. materials in the pesticide formulation that are not the active ingredient. Some inert ingredients may be toxic or hazardous to people.

inhalation. the method of entry of pesticides through the nose or mouth into the lungs.

inhibit. to prevent something from happening, such as a biochemical reaction within the tissues of a plant or animal.

insect growth regulator (IGR). a type of pesticide used for control of certain insects. Insect growth regulators disrupt the normal process of development from immature to mature life stages.

insecticide. a pesticide used for the control of insects. Some insecticides are also labeled for control of ticks, mites, spiders, and other arthropods.

inspection. the thorough checking of items for the presence of pests or pest eggs before bringing the items into a pest-free area.

integrated pest management (IPM). a pest management program that uses life history information and extensive monitoring to understand a pest and its potential for causing economic damage. Control is achieved through multiple approaches including prevention, cultural practices, pesticide applications, exclusion, natural enemies, and host resistance. The goal is to achieve long-term suppression of target pests with minimal impact on nontarget organisms and the environment.

interactive effect. interaction when two or more pesticides are mixed, producing greater or lesser toxicity to the target pests or changing the mode of action.

invertebrate. any animal having an external skeleton or shell, such as insects, spiders, mites, worms, nematodes, and snails and slugs.

knock down. an insecticide that has a rapid, although sometimes temporary, immobilizing effect on target insects; some knock down materials have rapid killing abilities.

larva. the immature form of insects that undergo metamorphosis (plural: larvae).

LC_{50}. the lethal concentration of a pesticide in the air or in a body of water that will kill half of a test animal population. LC_{50} values are given in micrograms per milliliter of air or water.

LD_{50}. the lethal dose of a pesticide that will kill half of a test animal population. LD_{50} values are given in milligrams per kilogram of test animal body weight.

leaching. the process by which some pesticides move down through the soil, usually by being dissolved in water, with the possibility of reaching groundwater.

legal threshold. a mandate to begin control of a particular pest. A legal threshold is usually based on a very low pest population and sets limits on the amount of pest damage or contamination allowed in food products offered for sale or endangering public buildings.

lethal. capable of causing death.

material safety data sheet (MSDS). an information sheet provided by a pesticide manufacturer describing chemical qualities, hazards, safety precautions, and emergency procedures to be followed in case of a spill, fire, or other emergency.

metabolism. the total chemical process that takes place in a living organism to utilize food and manage wastes, provide for growth and reproduction, and accomplish all other life functions.

metamorphosis. the physical transformation, more or less sudden, undergone by insects (and some other animals) during their development; the change of an insect from larva to pupa to adult.

microbial pesticide. pertaining to pesticides that consist of bacteria, fungi, or viruses used for control of weeds, invertebrates, or (rarely) vertebrates.

microorganism. an organism of microscopic size, such as a bacterium, virus, fungus, viroid, or mycoplasma.

mimic. relating to insect pheromones, the ability of a synthetic chemical to produce the same or similar effect on a target insect as a pheromone produced by that species of insect.

mode of action. the way a pesticide reacts with a pest organism to destroy it.

monitoring. the process of carefully watching the activities, growth, and development of pest organisms over a period of time, often utilizing very specific procedures.

MSDS. material safety data sheet.

multiple catch trap. a special type of trap designed to catch mice. A spring-loaded mechanism flips the mouse into a holding chamber; traps of this type can hold several mice. Mice are not killed by this device.

mutagenic. a chemical that is capable of causing deformities in living organisms.

mycelium. the vegetative body of a fungus, consisting of a mass of slender filaments called hyphae (plural: mycelia).

natural enemy. an organism that causes premature death of a pest organism; includes predators, pathogens, parasites, and competitors.

nonselective. a pesticide that has an action against many species of pests rather than just a few.

nontarget organism. animals or plants within a pesticide treated area that are not intended to be controlled by the pesticide application.

ocular. pertaining to the eye—this is one of the routes of entry of pesticides into the body.

ootheca. a capsule, constructed by female cockroaches, into which they deposit many eggs; some species carry an ootheca attached to the body, while others will deposit the ootheca in a hidden place. (plural: oothecae).

oral. through the mouth—this is one of the routes of entry of pesticides into the body.

organism. any living thing.

parasite. a plant or animal that derives all its nutrients from another organism. Parasites often attach themselves to their host or invade the host's tissues. Parasitism may result in injury or death of the host.

pathogen. a microorganism that causes a disease.

penetrate. to pass through a surface such as skin, protective clothing, plant cuticle, or insect cuticle. Also refers to the ability of an applied spray to pass through dense foliage.

perennial. a plant that lives longer than two years—some may live indefinitely. Some perennial plants lose their leaves and become dormant during winter; others may die back and resprout from underground root structures each year. The evergreens are perennial plants that do not die back or become dormant.

persistent pesticide. a pesticide that remains active in the environment for long periods of time because it is not easily broken down by microorganisms or environmental factors.

pesticide. any substance or mixture of substances intended for preventing, destroying, repelling, or mitigating any insects, rodents, nematodes, fungi, or weeds, or any other forms of life declared to be pests; and any other substance or mixture of substances intended for use as a plant regulator, defoliant, or desiccant.

pesticide formulation. the pesticide as it comes from its original container, consisting of the active ingredient blended with inert materials.

pesticide resistance. genetic qualities of a pest population that enable individuals to resist the effects of certain types of pesticides that are toxic to other members of that species.

pheromone. a chemical produced by an animal to attract other animals of the same species.

photosynthesis. the process by which plants convert sunlight into energy.

physiological. pertaining to the functions and activities of living tissues.

phytotoxic. injurious to plants.

plant growth regulator (PGR). a pesticide used to regulate or alter the normal growth of plants or development of plant parts.

postemergent. an herbicide applied after emergence of a specified weed or crop.

potency. pertaining to the toxicity of a pesticide.

powder. a finely ground dust containing active ingredient and inert materials. This powder is mixed with water before application as a liquid spray.

prebaiting. placing nontoxic bait in a trap to overcome bait or trap shyness on the part of the target pest; once the target pest becomes used to feeding from the trap, the nontoxic bait is replaced with toxic bait.

preemergent. the action of an herbicide that controls specified weeds as they sprout from seeds before they push through the soil surface.

prepupa. in insects having complete metamorphosis, the resting life stage between larval and adult forms.

pyrethrins. the active ingredients of pyrethrum insecticides.

pyrethroid. a synthetic pesticide that mimics pyrethrin, a botanical pesticide derived from certain species of chrysanthemum flowers.

rate. the quantity or volume of liquid spray, dust, or granules that is applied to an area over a specified period of time.

repellent. a pesticide used to keep target pests away from a treated area by saturating the area with an odor that is disagreeable to the pest.

residue. traces of pesticide that remain on treated surfaces after a period of time.

restricted-use pesticide. a pesticide that can only be used by or under the supervision of a licensed or certified pesticide applicator.

resistance. see pesticide resistance or host resistance.

rodenticide. a pesticide used for control of rats, mice, gophers, squirrels, and other rodents.

runoff. the liquid spray material that drips from the foliage of treated plants or from other treated surfaces. Also the rainwater or irrigation water that leaves an area—this water may contain trace amounts of pesticide.

sanitation. a pest management practice that involves removing desirable food and habitat that could be used by and promote particular pests.

secondary pest. an organism that becomes a pest only after a natural enemy, competitor, or primary pest has been eliminated through some type of pest control method.

selective pesticide. a pesticide that has a mode of action against only a single or small number of pest species.

service container. any container designed to hold concentrate or diluted pesticide mixtures, including the sprayer tank, but not the original pesticide container.

signal word. the word "Danger," "Warning," or "Caution": that appears on a pesticide label that signifies how toxic the pesticide is and what toxicity category it belongs to.

site of action. the location within the tissues of the target organism where a pesticide acts.

soluble. a material that dissolves completely in a liquid.

soluble powder. a pesticide formulation where the active ingredient and all inert ingredients completely dissolve in water to form a true solution.

solution. a liquid that contains dissolved substances, such as a soluble pesticide.

solvent. a liquid capable of dissolving certain chemicals.

sorptive dust (or powder). a fine powder used to destroy arthropods by removing the protective wax coating that prevents water loss.

spot treatment. a method of applying pesticides only in small, localized areas where pests congregate rather than treating a larger, general area.

spring trap. a spring-loaded trap used to capture mice and rats.

stomach poison. a pesticide that kills target animals who ingest it.

structural pest. a pest such as a termite or wood rot fungus that destroys structural wood in buildings.

Structural Pest Control Board (SPCB). a division of the California Department of Consumer Affairs having the responsibility for regulating pest control for hire in commercial and residential structures and for certifying the qualifications of persons working in this industry.

suppress. to lower the level of a pest population.

surface water. water found in ponds, lakes, reservoirs, streams, and rivers.

suspension. fine particles of solid material distributed evenly throughout a liquid such as water or oil.

symptom. a sign which indicates the presence of a disease or disorder.

systemic pesticide. a pesticide that is taken up into the tissues of the organism and transported to other locations where it will affect pests.

target. either the pest that is being controlled or surfaces within an area that the pest will contact.

teratogenic. a chemical that is capable of causing birth deformities.

threshold. an established point in time, largely based on the size of a pest population, when control measures must be started to prevent health or safety damage or economic loss.

tolerance. the ability to endure the effects of a pesticide or pest without exhibiting adverse effects.

total release. a pressurized insecticide dispenser that will release its entire contents into an area once it has been triggered.

toxicity. the potential a pesticide has for causing harm.

toxic tracking powder. tracking powder that contains a poisonous material that can be absorbed through the skin or outer body covering of pests or ingested through grooming.

tracking powder. a fine powder that is dusted over a surface to detect or control certain pests such as cockroaches or rodents. For control, the inert powder is combined with a pesticide; the animal ingests this powder and becomes poisoned when it cleans itself.

translocate. the movement of pesticides form one location to another within the tissues of a plant.

ultraviolet. pertaining to light having a wavelength just beyond the violet end of the visible spectrum; such light is invisible to people, hence it is known as black light.

vaporize. to transform from a spray of droplets to a foglike vapor or gas.

vertebrate. the group of animals that have an internal skeleton and segmented spine, such as fish, birds, reptiles, and mammals.

volatile. able to pass from liquid into a gaseous stage readily at low temperatures.

water-soluble concentrate. a liquid pesticide formulation that dissolves in water to form a true solution.

wettable powder. a type of pesticide formulation consisting of an active ingredient that will not dissolve in water combined with a mineral clay and other inert ingredients and ground into a fine powder.

Index

CALIFORNIA REGIONAL POISON CONTROL CENTERS

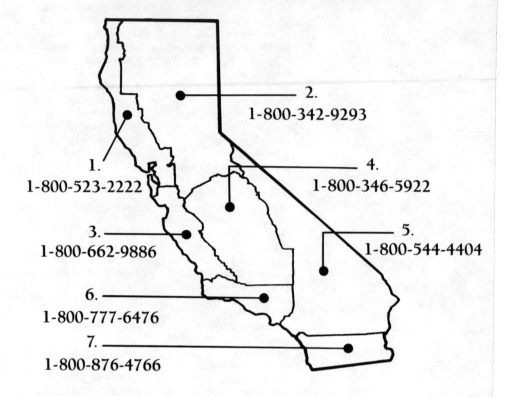

1. Del Norte, Humboldt, Mendocino, Sonoma, Napa,
   Marin, San Francisco, Contra Costa, Alameda,
   San Mateo counties: **1-800-523-2222**

2. Siskiyou, Modoc, Lassen, Shasta, Trinity, Tehama,
   Plumas, Butte, Glenn, Lake, Colusa, Sutter, Yuba,
   Sierra, Nevada, Placer, El Dorado, Sacramento, Yolo,
   Solano, San Joaquin, Amador, Alpine, Calaveras,
   Tuolumne, Stanislaus counties: **1-800-342-9293**

3. Santa Clara, Santa Cruz, San Benito, Monterey,
   San Luis Obispo counties: **1-800-662-9886**

4. Merced, Mariposa, Madera, Fresno, Kings, Tulare,
   Kern counties: **1-800-346-5922**

5. Mono, Inyo, San Bernardino, Riverside, Orange
   counties: **1-800-544-4404**

6. Santa Barbara, Ventura, Los Angeles counties:
   1-800-777-6476

7. San Diego, Imperial counties: **1-800-876-4766**